Contents

Introduction

Problem solving provides children with opportunities to embark on a learning adventure in which they can explore and investigate, take risks, make mistakes and come out the other end smiling, enthusiastic and eager to do more! It is about a process of challenge and not just about right and wrong answers.

Well-chosen problems provide pleasure and satisfaction. The activities in this book are planned to act as a magnet for enquiry, enabling us to access the children's ideas; to encourage them to puzzle; to prod their perceptions; to inspire dialogue and to activate mathematical thinking.

Who are they for?

The problems aim to be inclusive and so can be accessed by a wide range of ages and abilities. They are intended to be collaborative challenges and so are better suited to whole-class lessons, focus group sessions or maths buddy work. This allows children to pool their understanding, spark ideas from one to another, observe different ways of thinking and work through misconceptions together.

How do I use them?

The activities all follow a progressive structure but they aren't set in stone and so shouldn't be followed prescriptively. They are flexible enough for you to make them your own. They act as suggestions and outline possible routes you may wish to take. Inevitably you will need to draw on your experience and knowledge of your own class to fine-tune activities.

What process do I follow?

In order to teach children problem-solving skills they need a course of action to follow each time. The TEAR (Think, Explore, Act, Reassess) approach is a straightforward method that allows children to get to grips with any problem.

Think

- Hold back and ponder the problem.
- Refrain from rushing into an instant answer.
- Allow the problem to sink in and chew it over.

Explore

- Read and re-read the problem to make sense of it.
- Confer with others and discuss possibilities.
- Get to the root of the problem.
- Recognise key words, relevant information and superfluous information.
- Make informed estimates.

Act

- Consider options.
- Relate to other problems.
- Analyse relationships.
- Consult, negotiate and argue.
- Select resources.
- Write down ideas.
- Be systematic in approach.
- Keep good recording systems.
- Formulate and test hypotheses.
- Adjust approaches.
- Eliminate paths.

Reassess

- Scrutinise and evaluate efficiency of method.
- Ask: *Can the method be RIPEned (Refined, Improved, Polished, Edited)?*
- Communicate and report.
- Consider other ways of working.
- Extend and generate self-styled problems.

If children follow this TEAR method it builds an internal monitor and establishes a sound and business-like way of working.

What is my role?

The teacher's role, using these challenges, is to act as a facilitator to scaffold thinking processes, to promote reflection, offer feedback and engage the children in active learning where they can find their own ways of working.

Although you will provide the overall structure within a lesson, the lesson itself is organic and so is likely to contain many free-range moments that can't be planned for. The success of the lesson will depend on a commitment to low-level intervention and allowing the children to maintain ownership of their own learning. This involves leading and teaching from the perimeter but being at the centre of things at all times, observing, coaxing, chivvying and kindling unobtrusively. It's about making the classroom a think tank where narrowband and broadband thinking are nurtured.

Try not to provide children with instant answers or methods, but pose questions and 'food for thought' lines of enquiry. Motivate the children into searching for themselves and ensure learning experiences are peppered with surprises and frustrations to keep them on their toes. Remember to provide thinking time so that ideas can incubate and grow.

Above all, the role of the teacher is to have high expectations of everyone and to ensure that a sense of achievement is felt throughout the class.

What is the role of the learner?

It is important for children to realise that their part in their learning process is an active one and that means making decisions for themselves about what resources and equipment they need to use to solve problems. Ofsted (2002) recommend we build competence and confidence by making suggestions, such as 'would a matrix be of use when solving this problem?'; or by making available carefully selected equipment from which the children can make a reasoned choice. Of equal importance is allowing the children to find their own ways of presenting work independently.

How can the activities be used for assessment?

Formative assessment is at the heart of effective teaching and learning and discussion-based activities provide golden opportunities to gain access into children's mathematical thinking, allowing active assessment to take place. Children's explanations provide you with openings to assess diagnostically their level of understanding and allow you to intervene and plan accordingly. All of the problem-solving activities in this book can be used as dynamic assessment opportunities in order to maximise speaking and learning.

References

Thinking Maths: The Programme for Accelerated Learning in Mathematics, Adhami, M, Johnson, D and Shayer, M (Heinemann).

Thinking Things Through: Problem Solving in Mathematics, Burton, L (Basil Blackwell).

Active Assessment, Naylor, S, Keogh, B and Goldsworthy, A (David Fulton in association with Millgate House).

'Ofsted invitation conference for primary teachers', Brunel University, 20 March 2002. Ofsted (2002). (A conference designed to share best practice in problem solving, communication and reasoning in primary mathematics.)

Using and Applying Mathematics at Key Stage 2, Sellars, E and Lowndes, S (David Fulton).

Badger Maths Problem Solving: Skills and Strategies for Practical Problem Solving, Shapiro, S (Badger Publishing).

NRICH website: www.nrich.org

Chapter One

Finding all the possibilities

Finding all the possibilities is an open-ended strategy that involves children in searching for more than one solution to a problem. It requires the children to be flexible and inventive in their thinking, trying out different routes and paths to reach the same end point. This is extremely valuable practice, because children begin to understand that there can be more ways than one to crack a nut.

Searching for more than one answer demonstrates that while the outcome of a problem is the desired end result, it is the process and procedures that the children adopt to get there that matter. Children need to see this for themselves, and an analogy, such as a shopping analogy, can help them to understand. For example, Mrs Green goes to the shops every day, but how does she get there? Her destination is the same, but there is more than one way of getting there: walking, bicycling, bus, car, taxi, train, underground, plane and so on. The shops are the same shops at the end of the journey, but her method of transport to get her there could vary a great deal. Some systems of transport might be more reliable than others. Some might take longer. Explain to the children that it is the same with problems.

'Finding all possibilities' is about going on a maths safari in search of some treasure. There are likely to be problems and obstacles that test and frustrate along the way, but it's about persevering and searching for ways through the undergrowth, while confronting uncertainties head on. Above all, it is making children comfortable with the idea of taking risks in their work. They need to feel secure enough to know that if something doesn't work out, then it needn't lead to tears.

Magic threes

Setting the context

Four acute-angled trigons met each other three times a day in the number park to play a game of 'Magic threes' – a new maths team game that has got everyone thinking in Trigon City. It's called Magic Threes because the inventor, Mr Trio Thrice, believes that you can make any number from 1 to 10 using four number threes. For example, to make the number three itself you do the following:
$(3 \times 3) - (3 + 3) = 3$

The 'Magic threes' competition has gripped everyone in the city: 'Anyone that can make all the numbers 1–10 will win £3333!'.

As you can imagine, everyone is going number crazy!

Problem

Can you make the numbers 1–10 using four number threes and any operation?

Objectives

To solve mathematical problems and puzzles. To choose and use appropriate number operations and appropriate ways of calculating (mental, mental with jottings, pencil and paper) to solve problems.

You will need

Paper and pens; photocopiable page 24.

Preparation

Have the solutions ready.

Solving the problem

- Read out the story and the problem. Ask some questions to start the children thinking. Ask: *What other name is given to an acute-angled triangle?* (Equilateral or equiangular.) *If four acute triangles are joined together, what shape do they make?* (Irregular hexagon.) *What are the properties of the number three?* (The first odd prime number, second triangular number, fourth Fibonacci number.) *What does 'thrice' mean?* (Three times.) *Can you find the digital root of four threes?* (3, because $3 \times 4 = 12$, $1 + 2 = 3$.)
- Ask the children to work with a maths buddy to define the job of a pair of brackets as used in the maths sentence in the story. Establish that brackets show items that are treated together so that there is no confusion when performing operations.
- Challenge the class to make the number one with four threes. Remind the children that they can use an operation more than once and that brackets may be needed.
- After a few moments invite a volunteer to demonstrate. Make sure that the children understand that the line separating two numbers, such as 3/3, means to divide.
- Challenge the children to find the rest of the numbers from 1–10 using photocopiable page 24 to record their work. Explain that this can be done in any order, scoring the equivalent number of points to the number made (three points for the number three and so on). The children may work in pairs or on their own.
- Explain that the children will earn bonus points if they can name the number they make and list its factors. For example, 3 is an odd prime, it has two factors – 1 and itself.
- Now think about whether a number can be made in more than one way.
- Ask children to share their methods with each other before going through each number as a class:

(3 ÷ 3) + (3 - 3) = 1
(3 ÷ 3) + (3 ÷ 3) = 2
(3 × 3) – (3 + 3) = 3
$\frac{3 + (3 \times 3)}{3} = 4$
3 + 3 – (3 ÷ 3) = 5
$\frac{3 \times (3 + 3)}{3} = 6$
3 + 3 + (3 ÷ 3) = 7
(3 × 3) – (3 ÷ 3) = 8
(3 × 3) + (3 – 3) = 9
(3 × 3) + (3 ÷ 3) = 10

- Talk through each calculation together. Were some more difficult than others?

Drawing together

- Invite the children to look back at their methods. Do some numbers have more than one solution? See who scored the most points.
- Can the children make any of the numbers from 11 to 15 using just four threes?
- Consider whether the numbers 1 to 10 can be made using any other four numbers, such as four fives.

Support

Increase the number of threes so that children have more numbers to play with. For example, use six threes to make the number one:
3 ÷ 3 = 1
3 ÷ 3 = 1
3 ÷ 3 = 1
1 × 1 × 1 = 1

Extension

Set some new challenges along the same lines such as: *Join three sixes to make two.* For example: (6 + 6) ÷ 6 = 2. Alternatively, revise some of the examples looked at in the Plenary. Can the children think of other ways to get the same results?

Further idea

Play the dozen game: make 12 using any five numbers. For example:
2, 3, 4, 5, 6
(4 × 2 = 8; 8 + 3 = 11;
6 – 5 = 1; 11 + 1 = 12).

Quadrant

Setting the context

It's day two of the summer holidays and already Jimmy is getting restless. 'Mum, I'm bored.'

'Well, find something to do Jimmy. Granny and Grandpa are coming over later.'

'But that won't be for ages. What can I do?'

'Well you can clean your room for a start, instead of looking out of that window all day.'

'Actually the window is quite interesting. In fact I think I've got an idea! Thanks, Mum!'

Jimmy raced upstairs to find his art matchsticks and made the shape of the window with them. 'Mum, come here! I've got a problem for you!'

'Jimmy, *you* are a problem. Now what is it?'

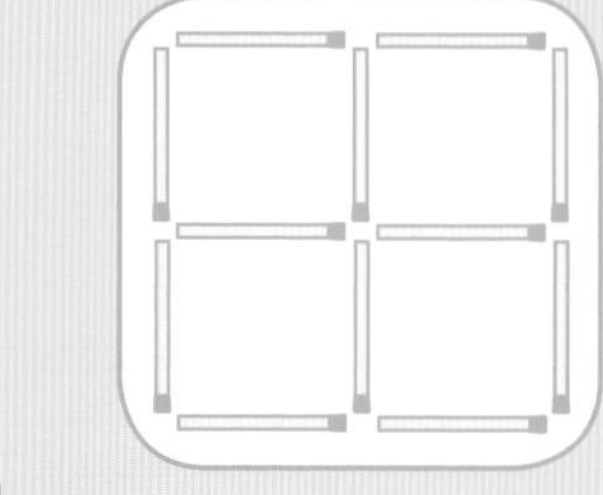

Problem

Can you move two matchsticks to make seven squares?

Objectives

To solve mathematical problems and puzzles. To recognise and explain patterns and relationships.

You will need

Headless matchsticks, straws, toothpicks or pencils; scissors; 12 rulers; maths dictionaries.

Preparation

Draw the solution to the above puzzle on the board and cover it.

Solving the problem

- Read the problem to the children. If appropriate, ask the children to act out the dialogue. Use the matchsticks as the starting point for setting other mathematical problems. Ask questions and set challenges such as: *How many quadrants can you see?* (There are five squares – four small, and one big square.) *Divide a gross by the total number of matchsticks, then add a score.* (144 ÷ 12 = 12; 12 add 20 is 32.) *Multiply the exterior by the interior and subtract from a right angle.* (8 × 4 = 32; 90 – 32 = 58.) *Subtract the total number of matchsticks from 1/4 of 10 squared, then state what type of number you have made.* (10 squared = 100; 1/4 of 100 = 25; 25 – 12 = 13; 13 is a prime number and a baker's dozen.)
- Provide each child with 12 matchsticks. Ask them to make the four-pane window shape while you take the opportunity to ask further questions: *How many squares are there?* (Five – four small and one big.) *How many right angles can you see?* (16 – four in each square.)
- Remind the children that there may be more than one solution, then set them off on the challenge to make seven squares.
- Provide the children with clues without giving them the answer, for example, *If you look at the window the solution is staring you in the face!*; *You'll be cross when you find out how it's done!*; *The squares don't have to be the same size.*
- Gather the children together and make the window shape using 12 rulers. Ask for ideas about how to tackle the problem. Move just one ruler into position and encourage the children to suggest where the other one might go.

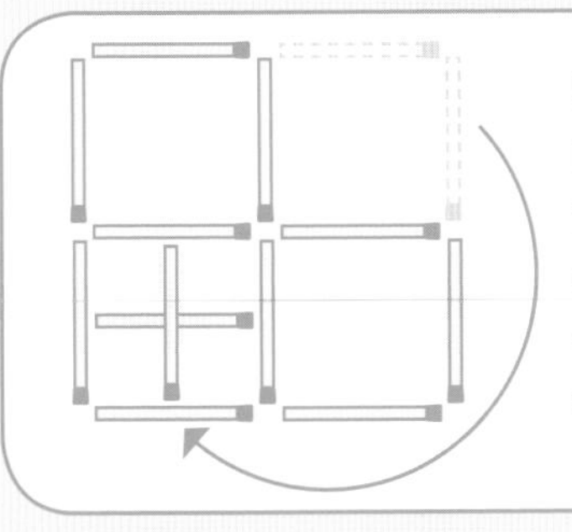

Move these two matchsticks: this makes seven squares (four small + three big).

- Show the children the cross shape inside another square and then count the matchsticks. Point out that the cross shape could be placed in any of the squares.
- Provide the children with clues to help them think about the problem at a deeper level. Define what a square is and then ask them to think about other shapes that might be called squares, such as a parallelogram and a rhombus.
- Now organise the children into larger groups for them to use maths dictionaries to research the names of other quadrilaterals. Do any of the definitions match the square?

Drawing together

- Look at the definition of a rectangle: *A quadrilateral with four right angles. Each pair of opposite sides is equal in length and parallel. The diagonals are also equal in length and bisect each other.* So, a rectangle is a square; therefore two matchsticks can be moved horizontally or vertically inside two different squares.
- Encourage the children to think beyond the conventional square shape. What other names can a square be called?

Support

- Start with four matchsticks and the instruction: *Add two matchsticks to make five squares*, before moving on to the main quadrant problem.
- Look at the following matchstick problem together:

Challenge the children to remove two matchsticks to make two triangles. Invite them to try and invent their own problem.

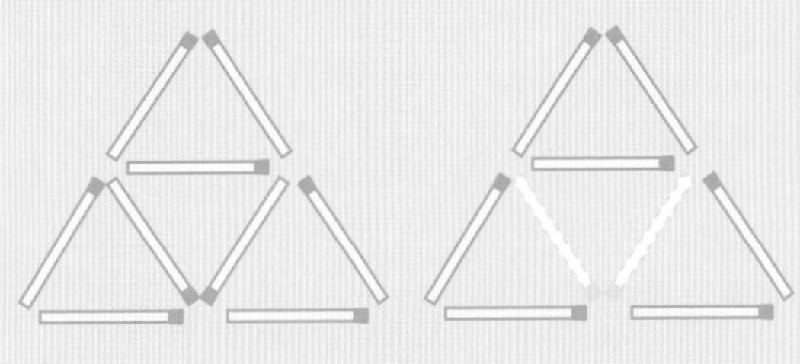

Dotted matchsticks show the two that are removed to leave two triangles (one big and one small).

Extension

Using the same window design, challenge the children to move four matchsticks to make three squares. (Move top-right horizontal and top-right vertical, move bottom-left vertical and bottom-left horizontal – then create three squares connected in a diagonal line.)

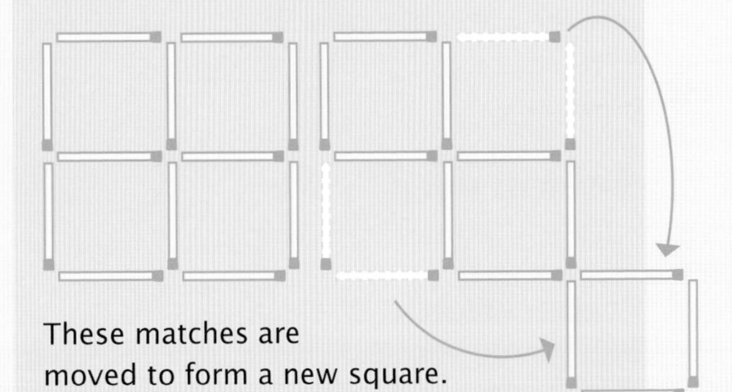

These matches are moved to form a new square.

Further idea

Challenge the children in mixed-ability maths pairs to make seven squares without making a cross shape.

Pooch police

Setting the context

Dear Scottie

The problem of dog congestion is getting worse. Since that new housing estate was built, the number of dogs on the field has quadrupled. There are so many dogs now that we all have to stay on our leads – we just walk around in one big line behind each other. It's like a dog traffic jam.

Anyway, I'm delighted that a new law has been passed so we can all enjoy the field again. Terriers can only go out on Monday, Wednesday and Friday and dogs bigger than terriers go out on Tuesday, Thursday and Saturday. Sunday is clean-up day. They've even got Pooch Police on the field to make sure that no one breaks the law. If you do you get a fine – you have to solve some problems. If you don't solve them then it's a night in the kennels.

Guess what? I got caught. I just nipped out for a little break (if you know what I mean) and the next thing I knew I was down the Pooch Station in a kennel. They gave me the 999 problem. I might be here for some time. Can you help me? I've had a go at one already:

537 + 462 = 999

Yours

Jack Russell

Problem

Make ten addition calculations with a total of 999, using the numbers 1–9.

Objectives

To solve mathematical problems and puzzles.
To recognise and explain patterns and relationships.

You will need

Paper and pens.

Preparation

Write out six solutions on the board, covered.

Solving the problem

- Read the story and the problem.
- Ask the children a number of questions related to the number nine to get them warmed up: *What is a nine-sided shape called?* (A nonagon or an enneagon.) *How many factors does nine have?* (Three – 1, 3 and 9.) *How many nines in 999?* (111) *Add ninety-nine and nine, then find the digital root.* (99 + 9 = 108; 1 + 0 + 8 = 9.) *How would you write the number nine in Roman numerals?* (IX) *How many seconds in nine minutes?* (540) *What do you get if you add the numbers from 1 to 9 together?* (45)
- Look at the addition Jack Russell has completed. Is it right? Can the children spot any patterns? (All the numbers on the top are odd and all those on the bottom are even.)
- Before moving on to other solutions, freeze-frame the addition and look at the different parts of the sum.
- Set the problem, awarding nine points for each addition found that equals 999. Can the children find more than six?
- Give children the time to work through some solutions before looking at the variations produced. Has anyone worked out

that numbers can be reversed? (234 + 765 = 999, becomes 432 + 567 = 999.) Set the children working again, challenging them to spot any other patterns.

- Ask the children whether two even numbers when added together can make nine. Can they find a rule for adding odd and even numbers? Draw a simple table to show variations that make nine (the complements of nine).

For example:

1	2	3	4	5	6	7	8
8	7	6	5	4	3	2	1
9	**9**	**9**	**9**	**9**	**9**	**9**	**9**

- Look at the digital root patterns made from the sums. For example, 142 + 857 = 999. 1 + 4 + 2 = **7**; 8 + 5 + 7 = 20; 2 + 0 = **2**: 7 + 2 = 9 so the digital roots of the numbers being added are 9 and the digital root of 999 is also 9. Does this help the children to recognise whether a combination of numbers has been chosen correctly?
- Call the children together and talk through some of the solutions found. Draw comparisons between some of the answers.
- Let the children join forces with another maths pair to brainstorm any further solutions.

Drawing together

Talk about all the variations that can be found and the patterns made. Ask the children to work out how many points have been made if each sum is worth nine points.

Support

Start off by finding all the variations for 555 then 666 before moving on to 888 and 999. Encourage them to draw number complement tables, then note the digital roots for each number. Are there any patterns?

Extension

- Ask the children how to write 999 in Roman numerals. (CMXCIX)
- Challenge the children to use the numbers 1, 3 and 5 to make three three-digit numbers to make 999 (153 + 315 + 531 = 999, 351 + 135 + 513 = 999).

Further ideas

- Try to make 1000 using the digits 0–9.
- Try to make the number 1089 by using any three digits, reversing them and then adding. For example: *469 + 964 = 1433. Close, but not close enough! Who can make the number exactly?*

Mr Ram the mind reader

Setting the context

The fair was back in town and the new attraction was causing quite a stir. People swarmed around Mr Ram the mind reader, like bees round a honey pot.

'Roll up, roll up!' shouted the attendant, 'Marvel at Mr Ram's incredible mind. Dare you have a go?!'

Joey had a go. He read the instructions at the counter:

> Look at the six number boards. Now choose a number, but don't tell Mr Ram. Just point to the board it appears on. If it appears on more than one board then point to those boards too. Mr Ram will tell you your number.

Joey pointed to boards 1, 2, 3 and 5. (See photocopiable page 25.)

Mr Ram looked Joey in the eyes for about ten seconds and then said, 'Your number is ... 29!'

Joey couldn't believe it. It was 29.

Problem

How did Mr Ram know that Joey's number was number 29?

Objectives

To solve mathematical problems and puzzles.
To recognise and explain patterns and relationships.

You will need

An enlarged copy of photocopiable page 25.

Solving the problem

- Read out the story and present the problem.
- First, ask a range of questions about numbers on the boards. For example, on board 1: *What do you get if you add the first row with the number second from bottom in column 3?* (36 + 41 = 77.) *Subtract the number of even numbers on board 4 from the odd numbers on board 1, then add this to the smallest prime number on board 3.* (30 – 15 = 15, 15 + 11 = 26.)
- Ask the children to look carefully at the boards that contain the number 29. Is there anything special about them? Do they share other numbers? (For example, 31 also appears on boards 1, 2, 3 and 5.)
- Ask the children to talk in small groups about the problem.
- Take on the role of Mr Ram and ask one of the children to choose a number and point to the boards it appears on. Hold back from telling the children the solution (for each of the boards pointed to, add the number in the top-right corner).
- Repeat this three or four times. Remember to exploit the number chosen by drawing out its properties. Then encourage the children to try the trick in pairs.
- It is unlikely that the children will be able to solve the problem without help so give them a clue: *If this problem has you cornered, then that's the place to start looking!*
- Share the solution so that the children can test different numbers.
- Re-enact the fairground scene, letting the children take the part of Mr Ram.

Drawing together

- Can the children work out why the method works? Does it work if the top-left numbers are added together?
- Share an example that does use the top-left numbers and ask the children to practise using the boards below.

Board 1

1	3	5	7	9	11	13	15
17	19	21	23	25	27	29	31
33	35	37	39	41	43	45	47
49	51	53	55	57	59	61	63

Board 2

2	3	6	7	10	11	14	15
18	19	22	23	26	27	30	31
34	35	38	39	42	43	46	47
50	51	54	55	58	59	62	63

Board 3

4	5	6	7	12	13	14	15
20	21	22	23	28	29	30	31
36	37	38	39	44	45	46	47
52	53	54	55	60	61	62	63

Board 4

8	9	10	11	12	13	14	15
24	25	26	27	28	29	30	31
40	41	42	43	44	45	46	47
56	57	58	59	60	61	62	63

Board 5

16	17	18	19	20	21	22	23
24	25	26	27	28	29	30	31
48	49	50	51	52	53	54	55
56	57	58	59	60	61	62	63

Board 6

32	33	34	35	36	37	38	39
40	41	42	43	44	45	46	47
48	49	50	51	52	53	54	55
56	57	58	59	60	61	62	63

Support

Start with the following number boards to give the children confidence. This trick requires adding the numbers in the top-left corner.

Board 1

1	3	5	7
9	11	13	15

Board 3

4	5	6	7
12	13	14	15

Board 2

2	3	6	7
10	11	14	15

Board 4

8	9	10	11
12	13	14	15

Extension

This is an outstanding trick that the children can practise on each other once they learn the method. Look at the number board below. With the board hidden from you, ask a child to select a circled number, then offer to tell them the seven-digit number underneath.

To reveal the seven-digit number, follow these steps:

1. Add 11 to the chosen number.
2. Reverse the result. Now start writing down the digits.
3. Keep on adding the two previous numbers, leaving out the tens.
4. Say the number.

For example, say that the chosen circled number is 12:

1. Add 11 to 12 to get 23.
2. Reverse 23 to get 32.
3. Add 3 and 2 to get 5; add 2 and 5 to get 7; add 5 and 7 to get 12 (omit 10 and just put down the 2); add 7 and 2 to get 9; add 2 and 9 to get 11, but omit the 10 and just put down 1.
4. The number is 3257291!

Col 1	Col 2	Col 3	Col 4	Col 5	Col 6	Col 7
(23)	(39)	(18)	(22)	(4)	(38)	(16)
4370774	0550550	9213471	3369549	5167303	9437077	7291011
(2)	(45)	(30)	(34)	(25)	(6)	(15)
3145943	6516730	1459437	5493257	6392134	7189763	6280886
(9)	(37)	(46)	(3)	(1)	(17)	(32)
0224606	8426842	7527965	4156178	2134718	8202246	3471897
(21)	(5)	(44)	(11)	(41)	(19)	(8)
2358314	6178538	5505505	2246066	2572910	0336954	0101123
(29)	(12)	(33)	(13)	(43)	(7)	(10)
0448202	3257291	4482022	4268426	4594370	8190998	1235831
(49)	(14)	(24)	(47)	(26)	(40)	(28)
0662808	5279651	5381909	8538190	7303369	1561785	9325729
(31)	(27)	(35)	(48)	(20)	(42)	(36)
2460662	8314594	6404482	9549325	1347189	3583145	7415617

Further idea

Choose any number, such as 35. Multiply by 3 = 105; add 2 = 107; multiply by 3 = 321. Add a number that is 2 more than the number first thought of. The number after the unit digit in the final answer will always be the number first thought of = 358.

Dear doctor

Setting the context

'Take a seat. What seems to be the problem?'

'I'm just not feeling very well.'

'How long have you been feeling like this?'

'Since I got my maths homework. I can't work it out. It's a really hard problem. I've been asked to write a dozen numbers between 69.2 and 69.3.'

'Ah, I see. You've been metagrobolised. Now I can't tell you the answer but I can write you a prescription. Take two prime numbers three times a day and talk about the problem with a friend. Two heads are better than one!'

Problem

Can you write 12 numbers between 69.2 and 69.3?

Objectives

To solve mathematical problems and puzzles.
To know what each digit in a number represents.
To multiply and divide whole numbers, then decimals, by 10, 100 or 1000.

You will need

Blank number lines; paper and pens.

Preparation

Draw decimal number lines on the board.

Solving the problem

- Read out the story and then set the problem to the class.
- Say that decimal numbers are called 'real numbers' or 'floating point numbers'. Ask children to tell you the place and face value of each of the numbers. For example, the place value of 6 in 69.2 is 60 and the face value is 6. What do the numbers after the decimal point mean? Whole numbers or parts of a number?
- Ask the children to: *express 69.2 and 69.3 as fractions* (69 2/10; 69 3/10); *find the digital root* (69.2 = 8, 69.3 = 9); *add the two together* (69.2 + 69.3 = 138.5); *subtract the smaller number from the larger and state what type of number is made* (69.3 – 69.2 = 0.1, which has the special name of 'tithe').
- Now begin the problem by asking the children to think of whole numbers. What are the numbers between 69 and 75? What other types of numbers can they think of that aren't whole numbers that may fit into a number line, such as decimals and fractions?
- Allow the children a few minutes to think about the numbers that fall between zero and one. Use an empty number line to show the progression of numbers in tenths and express these as decimals too.
- Using the number lines, ask the children to write 69 at one end and 70 at the other. What numbers go in between these?

Hopefully the children will soon recognise that 69.1–69.9 are all smaller than 70. Point out that 69.1 could be written with lots of other numbers after it that will still be smaller than 69.2, such as 69.13542. Show that this number has five decimal places.

- Help the children to think about place value names. Focus on the names to the right of the decimal point. Ask: *Is 69.2581470 bigger than 62.3 or smaller? How many decimal places does it have?* Many children will see it as being bigger because it is longer and contains numbers bigger than 3. Make it clear that the number is still smaller and it would only be bigger if the value of the tenths was to change to 3.
- Allow the children time to think about the number line and to write down any other numbers they can, without changing the value of the tenths.
- Discuss what types of numbers have been made and check that no one has changed the value of the tenths.

Drawing together

- Check that the children understand that there is an infinite amount of numbers in between numbers. Ask the children to tell you what number follows 1 on a number line. Children might say 2 or may give numbers such as 1.1, 1.2, 1.3489216 and so on.
- Repeat for other decimal numbers.

Support

Show the children that a number can have many digits after the decimal point but that the digits before the decimal point remain the only whole numbers. For example, 5.77777777777777 still means five whole ones and some parts. Ask: *Is this greater than or less than six?* The sevens and the number of them may confuse the children, so this may need reinforcing. Practise with smaller numbers, making full use of number lines and filling in an empty place value chart.

Extension

- Encourage the children to write down the numbers between large numbers such as 1,000,000 and 1,000,001. How many are there?
- What comes after ten-millionths on a place value chart? Can the children expand the values on both sides of the decimal point?

Further ideas

- Invite the children to roll a dice three times to make a decimal number, such as 21.1. They then make a new number using the same tens and units, but change the tenths digit, such as 21.2. Now ask them to write a score of numbers in between those two numbers.
- Ask the children to roll a dice four times to make ten decimal numbers. Challenge them to find the place and face value of each number and the digital root.

My head's swimming!

Setting the context

He'd been working on it for weeks and now it was finally done. Ian had made the swimming pool for his goldfish, 'They'll love it! It's a rectangular heaven. All I need now is some rope to go around the edge. I'll just give Elliott a ring.'

'Elliott, it's Ian. I've finished the pool. I want to order some rope from you to go round the edge.'

'Sure. What size do you need?'

'Well the pool area is 36m squared.'

'No problem, I'll cut four pieces and bring them over.'

Later that day the rope arrived. Ian wasn't very happy. 'Hang on, the pieces of rope don't fit. They're the wrong size!'

Elliott was confused. 'But you said the area was 36 metres squared so I bought along two four-metre pieces and two nine-metre pieces, and 9 × 4 is 36!'

'Yes, 9 × 4 is 36, but my pool isn't nine metres by four metres. The sides have different measurements!'

'What? Oh, my head's swimming now!'

Problem

What might the length of the sides be?

Objectives

To solve mathematical problems and puzzles.
To measure and calculate the perimeter and area of simple shapes.

You will need

Photocopiable page 26; pens and pencils; rulers; calculators.

Preparation

Draw the different possible solutions on the board, covered.

Solving the problem

- Read the dialogue and present the problem to the children.
- Ensure that the children understand the distinction between perimeter and area. Establish that area means the surface of a shape or object. The area of a regular shape is found by multiplying its length by its width. Perimeter is the total length of the sides bounding an area.
- Ask the children what they think the perimeter of a 9m × 4m swimming pool would be (9 + 9 + 4 + 4 = 26m). Stress that the answer is not in square units. Elicit that Ian is giving the area, but it is the perimeter that Elliott needs to know.
- Allow pairs some time to think about what numbers, when multiplied together, make 36. Use the opportunity to discuss factors: a factor of a number is one of two or more numbers that are multiplied together to make the number.

- List the factors of 36 together: (1, 36) (2, 18) (3, 12) (4, 9) (6, 6) or 1, 2, 3, 4, 6, 9, 12, 18 and 36. Tell the children that 1 is always a factor of any number, as is the number itself. Explain that a prime number is a counting number that has only 1 and itself as factors. For example, 7 is a prime number because the only factors are 1 and 7. A composite number is a counting number that has factors other than 1 and itself. For example, 12 is a composite number because its factors are 1, 2, 3, 4, 6 and 12.
- Establish that although Elliott had brought two nine-metre lengths and two four-metre pieces of rope, this was only one possible way to reach 36m. Point out that a rectangle could actually be a square shape. Draw a 6m × 6m square on the board.
- Encourage the children to consider whether there are more than five solutions (based on the whole number pairs). Ask them to divide 36 by the whole numbers that aren't factors, such as 5, 7, 8, 10, 11... 20, 21, 22, 23 and so on. How many more ways are there now?
- Provide the children with an example involving a decimal such as 36 ÷ 2.5 = 14.4. If this is written another way as 14.4 × 2.5 = 36, they will be able to see that there are many ways of making an area of 36 metres squared. Ask the children to divide 36 by any decimal number less than 36 to craft even more solutions.

Drawing together

- Ask the children whether they would be able to find all the possible solutions. Why is it easier to stick with whole numbers?
- Ask the children to re-enact the dialogue, but ask what Elliott should find out *(What is the length and width of the pool?)* Challenge the children to give possible answers to each other for quick checking with calculators. Let them record their results on photocopiable page 26.

Support

Use a different-sized rectangle as an example, such as eight or twelve metres squared. Ask the children to draw their swimming pools.

Extension

Work with bigger areas, such as 95m squared. Move on to finding the possible perimeters for rectangles with areas that aren't whole numbers. Ask: *What could the measurements be for a swimming pool with an area of 68.7 squared metres?*

Further idea

The perimeter of a pool in the shape of a scalene triangle is 15cm. What could its sides measure?

Save the rhubarb!

Setting the context

'Right then, Fran, can you water the garden while we're on holiday? You see that monkey puzzle tree? Well he's thirsty and I always give him seven litres every day. Then there's my runner beans. They have four litres a day. The rhubarb has one litre a day – he's very particular you know, one litre exactly!'

'No problem, Grandad. Do you have a hose pipe?'

'I use watering cans. I've left them in the shed. One holds seven litres and the other holds four litres.'

'But what about a one-litre can, Grandad?'

'You don't need one.'

'But what about the rhubarb? How can I measure out one litre?'

'Well that's for you to work out, Fran! I'll give you some extra pocket money when we get back. Remember, I'm entering that rhubarb in the garden show competition.'

Problem

How can Fran give the rhubarb exactly one litre using the two watering cans?

Objectives

To solve mathematical problems and puzzles.
To use all four operations to solve word problems involving length, mass and capacity.

You will need

A seven-litre container and a four-litre container; pebbles; photocopiable page 27; pencils.

Preparation

Fill the four litre container with water.

Solving the problem

- Read the story and pose the problem with the two containers at hand.
- Give groups five minutes to jot down ideas on the photocopiable sheet, and then survey the responses. Note that there may be more than one way to solve the problem.
- Now tell the children that the watering cans are not calibrated so they cannot simply measure one litre. This removes one possible solution!
- After further discussions, tell the children that they are not allowed to estimate one litre because Grandad said the measurement had to be exact.
- Children may ask whether they can use another container to measure, but no other containers are available. Explain that they have to visualise the problem and use the two containers.
- Organise them into pairs or threes and give the children time to talk through their ideas. Remind them to write down possible solutions as they go along.
- Freeze the discussions and ask for volunteers to talk through their favoured solution.
- Demonstrate the method: pour the four litres into the seven-litre can, leaving three litres empty. Fill the four-litre can again. Pour this into the seven-litre can until that is filled to the top. As the seven-litre can could only

hold another three litres, there will be one litre left in the four-litre container – this is for the rhubarb.

- Now ask the children to recall and explain the method to their partners.

Drawing together

Talk about whether there is more than one way of doing the problem. Can it be done in fewer steps? For example, what would happen if the seven-litre can is filled first?

Support

Tell the story of the Thirsty Crow (below) and see if children can work out how to help.

One day, a thirsty crow found a jug with some water at the bottom of it. 'At last, water. I'm so thirsty!'

She put her beak into the jug but the water was so low she couldn't reach it. She bent over further, but still couldn't reach the water. *(Ask the children what she should do.)*

Then the crow decided to try to tip the jug on its side.

'The water will run out and then I'll be able to drink it.' She pushed at the jug with all her strength but it wouldn't move. *(Ask for any more ideas.)*

The crow was now very thirsty and very tired. Suddenly she had an idea. She picked up a pebble and dropped it into the jug. It made the water a little higher. She dropped in every pebble she could find. As each pebble went in, the water rose a little bit higher. At last the water reached the top of the jug and the crow could take a drink!

Sometimes, strength doesn't matter when you've got good thinking!

Let the children test the crow's method.

Extension

Given a five-litre jug, a three-litre jug, and an unlimited supply of water, how do you measure out exactly four litres? (Fill the five-litre jug. Fill the three-litre jug from the five-litre jug, leaving two litres. Empty the three-litre jug. Pour the two litres from the five-litre jug into the three-litre jug. Refill the five-litre jug and pour one litre from it to fill the three-litre jug. That leaves four litres in the five-litre jug.)

Further ideas

- A nine-litre container is full and both a three-litre and five-litre container are empty. Without using any other containers, divide the water into equal amounts.
- You have a 12-litre container of water and you want to give your friend six litres. Your friend only has a five-litre and an eight-litre container. How can you give your friend six litres?

Hamster chute!

Setting the context

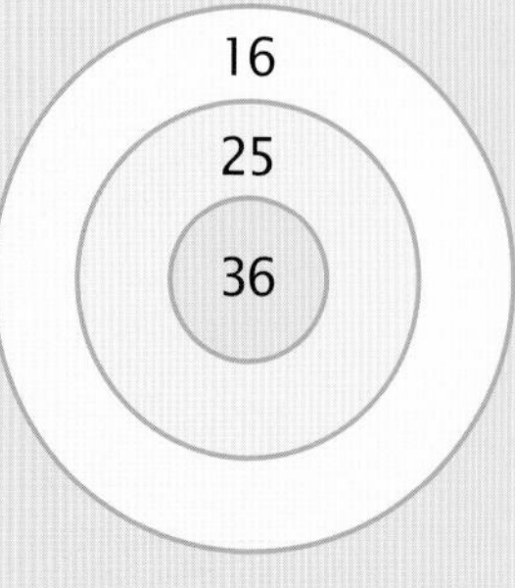

The British team were waiting by the door ready to jump. They could see the target way down below and were ready to go, 'OK lads, this is it. We're going to win this competition and win it in style. Just aim for the 36!'

The World Hamster Parachuting Competition was held every year in the Mediterranean. The target was a circular pontoon divided into three rings. Each ring was clearly marked with a number. Some years the numbers were high and some years they were low. If you missed the pontoon it meant that you ended up in the sea which was never pleasant because no one really enjoyed a mouthful of salt water.

'Right, hamsterchuters. On three. One, two, threeeeeeeeeee!'

The British team all landed on the pontoon, to the huge relief of their fans. Last year they all missed. They scored 36, 16 and 16 giving them a total of 68 points.

Problem

If all three hamsters land on the pontoon, what are their possible total scores?

Objectives

To solve mathematical problems and puzzles.
To recognise and explain patterns and relationships.

You will need

Photocopiable page 28; paper and pens.

Preparation

Draw three concentric circles on the board, labelling the rings 16, 25 and 36.

Solving the problem

- Read out the story, and then ask the children what they notice about the numbers 16, 25 and 36. They are all square numbers. Define a square number as a number raised to the second power (multiplied by itself).
- Go through the first ten square numbers together: 1, 4, 9, 16, 25, 36, 49, 64, 81, 100. Challenge the children to say them all in under ten seconds.
- Look at the arrangement of rings on the pontoon target. Ask the children what name is given to this arrangement of circles. They are concentric circles, which means two or more circles with the same centre.
- Focus on the problem and ask the children to consider all the possible totals the hamsters could achieve. Provide copies of the photocopiable sheet, paper and pens for the children to use. At this stage, offer no help so that the children can work through their own methods and lines of enquiry.
- Share some of the children's responses with the class.
- Remind the children that their manner of working needs to be systematic to cover all possibilities.
- Ask the children to work out how many possible combinations of score there are. Will some totals be the same?
- Work through the solutions together and see if the children agree with you:

36	+	36	+	36	=	108
36	+	36	+	25	=	97
36	+	36	+	16	=	88
36	+	25	+	25	=	86
36	+	25	+	16	=	77
36	+	16	+	16	=	68
25	+	25	+	25	=	75
25	+	25	+	16	=	66
25	+	16	+	16	=	57
16	+	16	+	16	=	48

- What do all the totals make? (770)

- After two rounds of jumping this is how the teams had scored:

Team	Points
China	196
USA	163
UK	145
India	145
Japan	136
France	132
Sweden	114

- Using the table above as a starting point, ask the children to create their own table of results after three rounds and work out how the scores might have been made.

Drawing together

- Ask the children to work together and decide what score combinations could have been made to give these totals. Can any of the totals be made in more than one way? (145 could be made with 77 + 68 and 97 + 48.) Continue with questions such as: *How many points separate Sweden and China? Would three jumps in the outer ring be enough for France to take top spot?*
- Do three numbers always give ten possible totals?

Support

Ask the children to look at the different combinations made by 6, 8 and 10. How many combinations make the same number, such as: 10 + 8 + 6 = 24 and 8 + 8 + 8 = 24. Build from here to include larger numbers.

Extension

Challenge the children to work out the combinations for the next three square numbers, or use much larger numbers for the three circles. From here, increase the number of concentric circles to four, then five and six. For example, can the children work out all the possible solutions for a pontoon with 13, 21, 34 and 55?

Further ideas

- Ask the children to think about changing the rules. For example, if a hamster landed in the water then this would be a deduction of nine points. If all three landed on the bullseye then that's a bonus of 49 points. Consider how these scenarios would affect the results.
- Challenge the children to think about using different number patterns for the pontoon scores, such as a Fibonacci or triangular pattern.

Magic threes

Hello, Mr Trio Thrice here. Do you like challenges? Well, try this little problem for size:

Can you make all the numbers from 1–10 using only four 3s and any combination of operations?

You score points for each number made. If you make the number 6 you get six points.

If you can write down the factors and name the number that's two more bonus points!

I'll give you a helping hand to get started:

$(3 \times 3) - (3 + 3) = 3$

Factors = 1, 3

Number name = prime

That's five points for me!

Now try to make numbers 1, 2, 4, 5, 6, 7, 8, 9 and 10. Good luck!

Number	Solution	Factors	Number name	Points
1				
2				
3	$(3 \times 3) - (3 + 3) = 3$	1, 3	prime	5
4				
5				
6				
7				
8				
9				
10				

Mr Ram – mind reader

Roll up, roll up!
Let the amazing Mr Ram read the numbers in your mind.
How he does it nobody knows.
Dare you accept the challenge?

Take a good look at the number boards.
Choose a number, but don't tell Mr Ram.
Just point to the board it appears on.
If it appears on more than one board point to those boards too.
Mr Ram will tell you your number and you will be amazed!

1

3	5	7	9	11	1
13	15	17	19	21	23
25	27	29	31	33	35
37	39	41	43	45	47
49	51	53	55	57	59

4

3	6	7	10	11	2
14	15	18	19	22	23
26	27	30	31	34	35
38	39	42	43	46	47
50	51	54	55	58	59

2

5	6	7	13	12	4
14	15	20	21	22	23
28	29	30	31	36	37
38	39	44	45	46	47
52	53	54	55	60	13

5

17	18	19	20	21	16
22	23	24	25	26	27
28	29	30	31	48	49
50	51	52	53	54	55
56	57	58	59	60	31

3

9	10	11	12	13	8
14	15	24	25	26	27
28	29	30	31	40	41
42	43	44	45	46	47
56	57	58	59	60	13

6

33	34	35	36	37	32
38	39	40	41	42	43
44	45	46	47	48	49
50	51	52	53	54	55
56	57	58	59	60	46

Once you have been told the secret of this trick, explore whether it works for every number.

My head's swimming!

Can you work out what lengths the sides of the pool could be?

Use this space to work it out.

Save the rhubarb!

Fran's Grandad has asked her to water his plants while he's away. He's left her a note to remind her.

Dear Fran

Just so you don't forget!

The monkey puzzle tree drinks 7 litres a day.

The runner beans drink 4 litres a day.

The prize rhubarb drinks 1 litre a day.

Thank you!

Love, Grandad xxx

Fran's Grandad gives her a 7-litre watering can and a 4-litre watering can to do the job. How can she do it?

7

4

Use this space to work it out.

Hamster chute!

Use the table to work out what the hamsters could score.

36	+	36	+	36	=	108
36	+	36	+	25	=	97
	+		+		=	
	+		+		=	
	+		+		=	
	+		+		=	
	+		+		=	
	+		+		=	
	+		+		=	
16	+	16	+	16	=	48

Chapter Two

Finding rules and describing patterns

Finding rules and describing patterns is one of the most commonly used tools for solving problems. This effective strategy is characterised by an orderly, planned and standardised way of working. Patterns are not always easy to spot and so finding them can be frustrating – which is why drawing a table or making a list can be so helpful. A table helps to display the data given so that connections between numbers become apparent and a solution becomes clear.

Although it is helpful to give children a helping hand and provide them with tables and their headings, you should encourage them to create their own visual representations first. This way the children are placed in the position of thinker and not receiver. It is important for children to see you respecting their ways of working. This strategy also relies on acting a problem out using concrete materials such as counters, blocks and even people. The beauty of this strategy is that it offers children a very practical way of understanding a problem and so is very well suited to kinaesthetic and visual learners. Acting out a problem is a superb way of plotting the movements required and so helps children to see a solution unfold before them.

When used together, the strategies of drawing a table and acting it out are excellent tools when tackling problems.

The root of the problem

Setting the context

Farmer Pong couldn't work it out. Someone was stealing his carrots. 'They must have come in the middle of the night. My hearing isn't what it used to be.' He even put extra scarecrows on duty but that wasn't enough to stop half of his crop from disappearing. 'I even stayed awake and watched out for them but my eyesight's on the blink too.'

The surprising thing was they hadn't really been stolen, just uprooted. You see, the carrots were found in the next field. Some in a pile and some in an interesting pattern. A triangular pattern. Next to the third pattern was an envelope. He opened the note and inside found a letter, covered in muddy fingerprints. It read:

Dear Pong
So, you've found us at last. What took you so long? Well, you're here now and so it's time to work. We've made a pretty pattern for you but can you carry on the pattern to make 25 triangles?

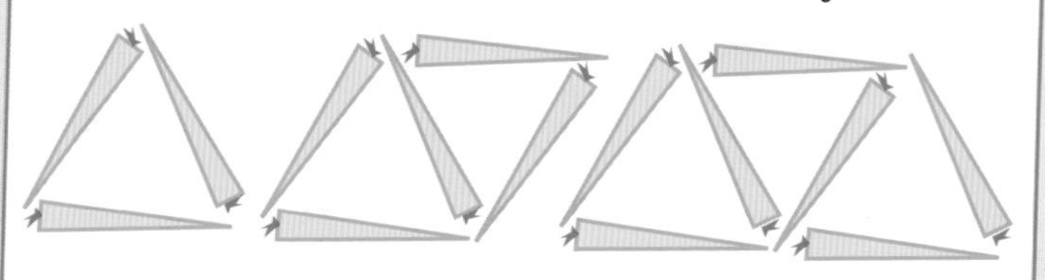

Hope you've eaten plenty of fish because you're going to need all the brain food you can!
The Carrot Crew

Farmer Pong enjoyed puzzles like this, so he set about trying to solve it.

Problem

Work out how many carrots are needed to make 25 triangles.

Objectives

To explain methods and reasoning, orally and in writing.
To solve mathematical problems or puzzles, recognise and explain patterns and relationships, generalise and predict.
To suggest extensions asking 'What if...?' questions.
To make and investigate a general statement about familiar numbers or shapes by finding examples that satisfy it.

You will need

A collection of headless matchsticks or lollipop sticks; photocopiable page 46; pencils.

Preparation

Draw the pattern on the board and a table of results ready to complete. Have ready enough matchsticks on each table.

Solving the problem

- Read out the story, and then set the problem.
- Ask the children questions relating to the triangles, such as: *Can you say what type of triangle the first shape is without using the word equilateral? If each carrot measures 45mm, what is the perimeter of the shape in centimetres? How many degrees will each interior angle be? What shape do two triangles joined together make?; How many*

carrots will there be in the fourth shape in the sequence?

- Discuss ways into the problem. Allow the children 'wait time' to think through their ideas.
- Ask the children to think of examples of different materials that could be used to act out the carrot problem (pencils, straws, pipe-cleaners and so on).
- Encourage the children to spot a relationship between the number of carrots and the number of triangles. Can they predict what shape five carrots will make?
- Give the children the opportunity to construct the triangular pattern using matchsticks.
- Encourage the children to describe a pattern according to the number of carrots needed.
- Lead the children into thinking about using a table to help them solve the problem systematically. Photocopiable page 46 offers one way of starting, but encourage the children to construct their own tables.

Drawing together

- Go through the problem together and refer to the table you have constructed on the board.

Number of triangles (t)	Number of carrots (c)
1	3
2	5
3	7
4	9

- Can the children work out that two more carrots are needed for each additional triangle?
- Prompt the children to write down, in their own words, a rule for finding the number of carrots. Share and discuss these rules.
- Generate a rule together using letters and numbers. A general rule for the number of carrots is $c = t \times 2 + 1$, or $c = 2t + 1$. To find how many carrots are made from 25 triangles, we say $25 \times 2 + 1 = 51$.
- Decide whether physically constructing the carrot shapes is actually necessary.
- Can the children work out the number of carrots needed for 36 triangles?

Support

- Give the children the opportunity to extend the patterns up to the tenth shape, helping them to construct a table of results. Encourage them to explain the pattern, then word a rule together.
- Invite the children to invent their own story, using sausages instead of carrots.

Extension

- Investigate other 'growing carrot' problems, such as using squares.

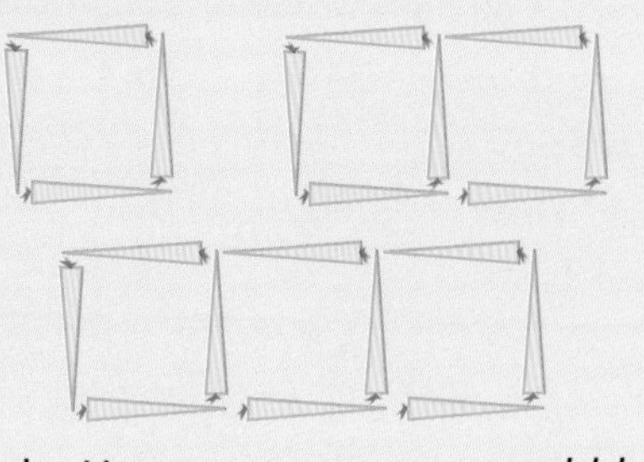

- Ask: *How many carrots would be needed to make 20/100/1000 squares?*
- Challenge the children to invent a similar problem for the Carrot Crew involving pentagonal or hexagonal patterns.

Further ideas

- Can the children write to the Carrot Crew setting them a new problem?
- Make a sequence using dots, such as the one below:

```
                      • • • • •
          • • • • •           •
• • •     •       •   •       •
• • •     • • • • •   • • • • •
```

Challenge the children to draw the next two shapes in the sequence and make a table. Decide which of the following is the correct rule: *The shape number times 4, then add 2*; *The shape number times 3 and then add 3*; or *The shape number times 2 and then add 4.*

A real handful

Setting the context

This letter arrived yesterday. It read:

> **Monkey Manner School Reunion**
>
> Dear ex-pupil,
> We are delighted to announce a reunion of your class at Monkey Manner School on 1 April. Please make a date for your diary!
> We look forward to seeing you all soon,
>
> Yours
> The Monkey Manner Teachers

The letter took me by surprise. I hadn't seen my old classmates for so long. It must have been ten years. I wondered if they had changed. What would they be doing now?

In the end, only nine of us turned up! There should have been 45. I couldn't understand why the others didn't come. Anyway, we were glad to see each other. In fact when we first met we just shook each other's hands for ages! We were a handful of cheeky monkeys at school and I suppose we still are. A leopard never changes its spots and a monkey never loses its cheek!

Problem

How many handshakes are there?

Objectives

To explain methods and reasoning, orally and in writing.
To solve mathematical problems or puzzles, recognise and explain patterns and relationships, generalise and predict.
To make and investigate a general statement about familiar numbers or shapes by finding examples that satisfy it.

You will need

Paper and pens; cubes.

Preparation

Draw a triangular number pattern on the board, such as the one shown here:

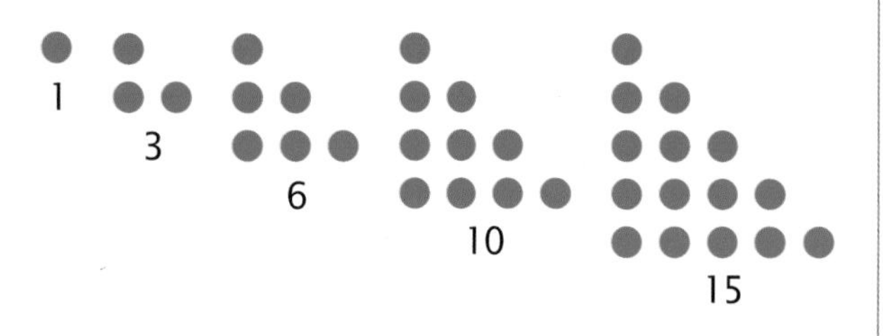

Draw a blank table ready for completion with results (see 'Drawing together' on p33).

Solving the problem

- Read the story aloud to the children, then set the problem.
- Together, think of ways to crack the problem. For example, act it out, make a list, look for a pattern. Record these on the board for the children to mull over and discuss with their neighbours.
- Begin acting out the problem by asking three children to shake hands with each other, role-playing the reunion.
- Now repeat this for four children and then five. Invite the rest of the class to count and record the number of handshakes each time. Ask: *How can this be recorded? Is it more sensible to remember it or to write it down?*
- Ask questions that help the children to concentrate on working systematically and looking for patterns. (Think of patterns that might emerge such as square, triangular, cube.) In groups, ask the children to think of ways of jotting down their solutions.
- As the children act out the problem, remind them to keep track of the number of handshakes. How will they do this?

- Ask the children to think about what they notice about the number of handshakes and the number of people.
- Share the results and ask for any insights that would prove helpful in solving the problem without having to act it out.

Drawing together

- Explain to the children that the maths they use in the problem depends on the approach used, but they should all come to the same answer.
- Find out if the children spotted that the problem involved triangular numbers. Refer to your table of results and the triangular pattern that can be generated from it. By how many do the triangular numbers jump up each time?

Number of monkeys	Number of handshakes
1	
2	1
3	3
4	6
5	10

- Did anyone fill in one handshake for one person?

Support

- Act out the problem in a group of four children. Record the pattern being made using cubes arranged as triangles.
- Encourage the children to construct further triangular patterns.

Extension

- Challenge the children to work out how many more monkeys turned up at the reunion if there was a total of 106 handshakes.
- Read this rule to the children: 'If you multiply the number of monkeys by one less than the number of monkeys, then divide by two you will always find the number of handshakes.' Does this work? Ask the children to investigate. Ask them to write this as a number expression, for example: $\frac{n\,(n-1)}{2}$
- How many handshakes are there at the reunion if three monkeys come with their partners and shake hands with everyone except their own partners? (Six monkeys = 15 handshakes: 15 – 3 = 12; subtract 3 for the shakes that would be between partners.)

Further ideas

- Investigate how many handshakes there would be if everyone in class shook hands twice with everyone else.
- Set this interesting problem. There are ten people in a room, all of different heights. If they shake hands only with people who are taller than them, how many handshakes take place in the room? (Zero!) Let's say that A and B want to shake hands. For A to shake hands with B, B must be taller. But for B to shake hands with A, A must be taller. This is a paradox, since both are different heights (and no two people can possibly be taller than each other) so no handshakes occur!

Fast fingers

Setting the context

It was an early summer afternoon and Filly was bored. She'd played football, ridden her bike, annoyed her brother and eaten all her crisps. Filly lay flat on the grass, looking up at the sky. She admired the patterns the clouds seemed to make, 'How beautiful! It looks like a necklace... Oh heck, next week!' she screamed.

Next week was pretty important. It was the Mumble County Daisy Chain Finals. She came second last year and third the year before. Was this going to be her year? The competition was tough. You had to link as many daisies as you could in just 25% of an hour. Yes, 25%.

She decided to challenge herself to see how many daisies she could thread minute by minute. Her brother agreed to time her and noted down her results. She was always slow to start because they were difficult to thread but she soon quickened up. She joined two daisies together in the first minute, three in the next minute, four in the third... incredible! They didn't call her Filly Fast Fingers for nothing.

Problem

How many daisies will there be in the chain after 25% of an hour?

Objectives

Solve mathematical problems or puzzles, recognise and explain patterns and relationships, generalise and predict. Recognise and extend number sequences, such as the sequence of square numbers, or the sequence of triangular numbers 1, 3, 6, 10, 15.

You will need

Paper-clips; watch or stopwatch; photocopiable page 47.

Preparation

Draw a completed table of Filly's results (see puzzle above) on the board, hidden at the start of the lesson.

Solving the problem

- Read out the story and the problem. Ask: *How much is 25% of an hour? What about 50%? 75%?*
- Invite pairs to freewheel their thoughts and ideas about how to solve the problem.
- Record the children's ideas and display them clearly on the board to give the children pride and confidence in their work and ideas.
- Work together to expand upon ideas that show systematic thinking.
- Now provide the children with some paper-clips to work through the problem. Watch to see if any children independently begin to record their results. Don't prompt them at this stage. Ask if acting it out is necessarily the most efficient method.

- After listening to suggestions, nudge the children towards using a table if this hasn't been considered already.
- Give out the photocopiable sheet and ask the children to work in groups to start completing the table.
- Now invite everyone to stand on their feet around the perimeter of the classroom. Read the story again and act out the problem so that people become daisy chains and start to link together. Encourage the children to join hands or link arms each new minute as directed by the story.
- Send the children back to working in groups to finish recording the results in the tables drawn. Encourage the children to work with their paper-clips again, and as they record their results ask them to spot a pattern.

Drawing together

- Look at the results together and identify any patterns that would help us to find out how many daisies Filly could add after 23 minutes.

Minute	1	2	3	4	5	6	7	8	9	10	11	12	13	14	15
Added daisies	2	3	4	5	6	7	8	9	10	11	12	13	14	15	16
Total	2	5	9	14	20	27	35	44	54	65	77	90	104	119	135

- If Filly continues to 23 minutes there will be 275 daisies in her chain. At half an hour she will have 464. After 45 minutes she will have 1034 daisies and after an hour she will have 1829 daisies.
- Remind the children of the importance of checking and re-checking results. Look over the table again to give the thumbs up.

Support

Rework the problem so that the children have to find how many daisies are added after five minutes, concentrating on the jumps in between the total amounts. Ask the children to spot how many it goes up by each time and build from here to incorporate their predictions for six minutes, seven minutes and so on, with and without the paper-clips as support.

Extension

Challenge the children to work out how many daisies Filly would have in her chain after 900 seconds, 1500 seconds and so on.

Further ideas

- Challenge the children to invent their own problem involving a sailor who has to link a number of chain links together ready for the World Anchor Lifting Championships.
- Look for other patterns in sequences of numbers, such as:

0 1 1 2 3 5 8 13 _ _

6 9 8 11 10 13 _ _

1 3 7 13 21 _ _

Pick your brains

Setting the context

The trees on the planet of Foop have been attracting tourists from all over the galaxy for years. They produce potpotpinker fruit, transparent rhombi with soft pink centres, packed full of Vitamin C. They ripen every month of the year. A Foop month has 45 days, and there are 28 months in a Foop year.

Jesto, a potpotpinker picker, began collecting fruit as they became ripe, ready for shipping to Earth. He picked one potpotpinker on the first day, two on the second, four on the third and eight on the fourth day. For every potpotpinker fruit Jesto picks, he is paid three durkins.

After one-third of a month, an emergency order was suddenly received at Potpot HQ from Mrs Joyner from London, England on Earth. Jesto immediately packed all the potpotpinkers he had picked ready to send to Mrs Joyner. Jesto counted how many he had picked just to double check and sent them by first class speed shuttle to Earth.

Problem

Pick your brains to work out how many Jesto will have collected in one-third of a Foop month.

Objectives

To solve mathematical problems or puzzles, recognise and explain patterns and relationships, generalise and predict.
To recognise and extend number sequences, such as the sequence of square numbers, or the sequence of triangular numbers 1, 3, 6, 10, 15.
To find some common multiples.

You will need

Rhombus shapes and parallelograms; cubes; paper and pens; calculators.

Preparation

Draw a table of results on the board, hidden. Provide cubes on each table.

Solving the problem

- Before you tell the story of Jesto the potpotpinker picker, hold up a rhombus and ask the children what kind of shape it is. Do they know any other names for the shape? Hold up a parallelogram and compare the two shapes. Are they the same? Which features are the same?
- Invite the children to talk to a maths buddy and try to define 'rhombus'. Then write out a definition on the board: a rhombus is a parallelogram with four equal sides but no right angles. Other terms for rhombus include 'lozenge' and 'diamond'.
- Now read the story and present the problem.
- Help the children to grasp the problem by suggesting that they omit or block out the superfluous information in the problem text. Define superfluous as extra, unnecessary or redundant information that we can put in the 'recycle bin'.
- Decide what is the important information to retain, then ask the question: *What do we need to find out?* Read the problem again to cement what is actually being asked.
- Before going any further, ask the children to work out one-third of a Foop month and therefore how many days' worth of fruit we are being asked to find.
- Allow the children time to work with cubes (as potpotpinkers) to get a feel for the problem. Establish whether this is the most efficient way of finding a solution. Can the problem be better completed using a pencil and paper method? (The children should

already be looking for patterns in the numbers.)

- Direct the children to work in pairs to find a solution by compiling a table from the information given. Give them time to collaborate and discuss their ideas.
- After the children have worked through their solution ask maths buddies to share their workings out with the class, discussing any problems they have encountered.
- Point out that this is a doubling pattern. Work through the solution on the board by completing the table, and ask the children to check how close they were to the solution. Insert place holders where there are spaces. Did anyone make some doubling errors that skewed the results? Look at the solution together and discuss where the children may have gone wrong.

Day	Potpotpinkers
1	00001
2	00002
3	00004
4	00008
5	00016
6	00032
7	00064
8	00128
9	00256
10	00512
11	01024
12	02048
13	04096
14	08192
15	16384

Drawing together

Suggest the following to the children: *To find how many potpotpinkers Jesto picked in 30 days, just double your answer for 15 days.* Is that right? Can the children explain their answer?

Support

Ask the children to find out how many potpotpinkers Jesto found on the eighth day. Provide calculators for doubling larger numbers.

Extension

- Capitalise on the numbers for a Foop month by generating some worthwhile puzzles and number problems. For example, how many days are there in a year on Foop? How many days in one-seventh of a year?
- Challenge the children to solve the following: Jesto picked one fruit on day one, three on day two, nine on day three, 27 on day four. How many did he pick on day 17? (43,046,721.) The children should see that the numbers are trebled.

Further ideas

- Work out how many durkins Jesto earns in a third of a Foop month. Change his rate of pay to different multiples and ask the children to calculate his earnings for 64 potpotpinkers.
- Challenge the children to invent their own problem. For example, *Ant likes to collect shells. He visits the beach every day. On day one he collects....*

Walkie-talkies

Setting the context

Two isosceles triangles, Alpha and Beta, decided to go on a two-week walking holiday across their school field. They loved walking and they loved talking too. Sometimes they would walk and talk at the same time and at other times they would stop walking and just talk. This meant that some days they managed to travel bigger distances than others.

On day one they travelled 100mm despite nattering quite a lot. On day two they chattered a lot less and covered 130mm. On day three they were chatting again and travelled 20mm less than day two. By day four they needed a rest from talking and managed to walk 140mm.

'Do you know, Beta,' said Alpha. 'I think we're walking in some sort of pattern here...'.

Problem

How many millimetres did the triangles walk on the last day of their holiday?

Objectives

To solve mathematical problems or puzzles, recognise and explain patterns and relationships, generalise and predict.
To recognise and extend number sequences, such as the sequence of square numbers, or the sequence of triangular numbers 1, 3, 6, 10, 15.

You will need

Isosceles triangles; blank number lines; paper and pens; photocopiable page 48; cards with +30 and –20 written on them.

Preparation

Draw up the answer chart in advance.

Solving the problem

- Hold up two isosceles triangles and tell the children that you are going to read a story about a holiday these triangles had. Before you tell the story, test the children's knowledge and understanding about isosceles triangles. In small groups, encourage them to discuss and define an isosceles triangle. Share definitions and establish that it is a triangle with two sides of equal length and two angles of equal size. Ask the children how many lines of symmetry an isosceles triangle has (one) and what its rotational order of symmetry is (none).
- Now read out the story and set the problem. Establish that day 14 is the last day of the holiday. After a few minutes thinking and discussing time, brainstorm as a class any possible links between the numbers. Is there a pattern?
- Ask: *How many centimetres are there in 100mm? 130mm?*
- Share ideas about the numbers given and then provide thinking and talking time about how to solve the problem.
- Ask the children to tell their maths buddies what information they are given in the problem text and what they need to find out. How will they record this without writing out the whole problem?
- Invite the children to tell you how they will map out the problem and how they will communicate a solution clearly. Are they aware of any pattern that will help them?
- Allow the children to work independently, with a maths buddy or in small groups to arrive at a solution. Keep a low profile, but intervene when appropriate.
- After the children have attempted the solution, provide them with empty number lines, pens and triangles to act out their thinking by showing the distances covered.

Will they spot that the pattern involves two operations +30 and –20?

- Go out to the playground or field to demonstrate the distances the walkers are taking. Ask two children to act as the isosceles triangles. Mark each movement by placing the card +30 or –20 on the ground. What can the children see is happening each time?
- Go back to the classroom and prompt the children to draw up a table to help them calculate the distances covered.
- Look over the solution together and see if everyone's numbers match.

Drawing together

- Share solutions to the problem. Did the class reach a consensus? Were there any discrepancies? Can these be explained?
- Can the children work out how far the triangles would have travelled on day 20, supposing they had extended their holiday? (220mm.)

Day	1	2	3	4	5	6	7	8	9	10	11	12	13	14
Distance travelled (mm)	100	130	110	140	120	150	130	160	140	170	150	180	160	190

Support

Provide the children with different numbers to work with. For example, on day one the triangles walked 5m, on day two they went 7m, 6m on day three and 7m on day five. How far did they travel on day nine? These jumps are a lot less as the pattern is +2, –1.

Extension

- Ask the children to work out the total distance covered by the two triangles in mm, cm and m.
- The triangles go on another holiday to another field later in the year. On day one they go 8.5cm, 7cm on day two, 10.5cm on day three and 9cm on day four. How far do they travel on day 12? Day 16? Day 25? The pattern goes –1.5 then +3.5. Challenge the children to invent their own problems based on this pattern.

Further ideas

- Change the units of measurement to metres and kilometres, and instead of days use hours or weeks.
- Ask the children to write their own measurement problem about a group of astronauts who travel through space, covering different distances on each day.

Flea bag

Setting the context

Jazz is a little Jack Russell with a very big problem. Well, in fact, he has lots of little problems: fleas!

'It's not as if I don't take care of myself, you know,' Jazz insists wearily. 'I eat all the right foods, I exercise every day with my good friend Barry the Boxer and I even go to a Pooch Health Farm every two months. It's really driving me crazy now though. I'm just one big flea bag. Those pesky pinheads are multiplying at a rapid rate!

I started out with just two fleas, then the following week I had eight. By the third week there were 20 of them bouncing around in my fur. Well, this is week four, and Barry has counted 38! I've got nearly as many as him! He's covered too. It's no good, I'm going to have to call Rentoblast! Last time, I had to wait for days before they could come to see me. They said that there were other dogs fighting the same problem. I'll just have to keep my paws crossed that they can make it soon.'

Problem

If Rentoblast are unavailable, how many fleas will Jazz have on his back by the eighth week, when he's due for his health farm visit?

Objectives

To solve mathematical problems or puzzles, recognise and explain patterns and relationships, generalise and predict.
To recognise and extend number sequences, such as the sequence of square numbers, or the sequence of triangular numbers 1, 3, 6, 10, 15.
To find some common multiples.

You will need

Times-tables squares; photocopiable page 49; paper and pens.

Preparation

Draw the results table on the board and cover it.

Solving the problem

- Introduce the story of Jazz and his fleas. Present the number problem.
- With the children's help, isolate the important information within the problem and assess the problem. Concentrate on the word 'multiplying' in particular.
- Focus on the numbers 2, 8, 20 and 38. Can the children make the number 24 using any operation and all of the numbers? For example, 38 – 20 = 18, 18 + 8 = 26, 26 – 2 = 24. Are there any other ways to make 24 with the same criteria?
- Go back to the numbers 2, 8, 20 and 38 and ask the children to 'read between the lines'. Can they spot a connection between the numbers? Point out that what is in

between the numbers is often a good place to start.

- Provide the children with further clues if necessary and hint at using the times-tables square as a mental prompt. For example, *Perhaps multiple patterns hold the key to ridding Jazz of his fleas!*.
- Organise the children into trigon groups (mixed-ability groups of three) and encourage them to engage in a triangular conversation about the problem. Point out that disagreements are healthy in maths, because a solution is likely to be found through discussion.
- Take feedback from the group discussions. Then ask the children to construct a table so that they can systematically work out the problem. Offer prompts and tips to children who are struggling to make a connection between the numbers.
- Come together as a class and share the completed tables. Value all efforts made and work through any misconceptions. Identify the pattern and the operation involved.

Week	1	2	3	4	5	6	7	8
Fleas	2	8	20	38	62	92	128	170

Drawing together

- Reach an agreement that the total number of fleas on Jazz's back by week eight was 170.
- Discuss whether the problem could have been solved using a different method. (For example, would a drawing method have helped?) How did talking about the problem with others help?

Support

Approach the problem within the same context, but using different flea numbers. For example, *I started out with just two fleas, then the next week I had five, then by week three I had eleven fleas. On week four I had 20 fleas.* How many fleas did Jazz have by week six? Offer some direct support using the times-tables squares.

Extension

Use multiple patterns of numbers above the six times-table. For example: on week one Barry the Boxer started off with three fleas, on week two he had 15, week three, 39 and by week four he had 75. How many fleas did Barry have by week eight?

Further ideas

- David had been eating oily fish for a few weeks and suddenly noticed that his brain was a lot more active than it used to be. In week one he managed just one original thought, by week two he sprouted 10 amazing thoughts, and by week three he had produced 28 mind-blowing ideas. By week four his head exploded with 55 bright ideas. If his ideas keep multiplying at the same rate, how many ideas will David have produced in the ninth week?
- Challenge the children to create a word problem involving a multiple pattern.

Face painting

Setting the context

The holiday season has just begun in Pegness and everyone is getting ready for a busy holiday. Ingrid works as a children's face painter during the summer and already has some customers at her stall on the pier, 'Wow, customers already! I'm normally a bit rusty after a break so I won't be able to paint many faces to start with.' Ingrid is normally slow to start painting but each and every day of the season she gets quicker with more practice. She also gets more customers each day. On the first day this summer she painted four children's faces, on the second day she painted six faces, on the third day she painted nine faces. Yesterday, the fourth day of the season, she painted thirteen faces. She just got quicker and quicker. So quick in fact, that people from all around stopped to watch her at work as she painted faces faster than a bee's wings! Pegness was a popular place and that was partly down to Ingrid herself and her amazing painting powers.

Problem

If Ingrid keeps getting quicker, how many faces will she have painted by the eleventh day of the summer season?

Objectives

To solve mathematical problems or puzzles, recognise and explain patterns and relationships, generalise and predict.
To recognise and extend number sequences, such as the sequence of square numbers, or the sequence of triangular numbers 1, 3, 6, 10, 15.

You will need

Photocopiable page 50; pens.

Preparation

Draw eleven circles on the board to represent each of the days mentioned in the problem.

Solving the problem

- Read the story and present the problem.
- Discuss together with the children the information that there is within the problem that they can use to construct a table.
- Draw two rows on the board, one headed 'Day' and the other 'Faces painted'.
- Choose five children to demonstrate the problem. Ask one child to role-play Ingrid the face painter, then ask the rest to queue up for the face painting. Ask the painter to pretend to paint the faces of the first four children.
- Begin filling in the table from this starting point, putting 1 in the Day column and 4 in the Faces painted column.

© Paul Doyle/Alamy

- Next, ask six other children to come up for their faces to be painted. Ask the first four to sit down. Fill in the table as the last face is painted. Remind the face painter that she should be getting quicker with each face.
- Continue the role-play for day three, asking nine children to come up for a makeover.
- Remind the children of the value of acting out a problem to illustrate what needs to be done.
- Now look at the table so far and encourage the children to see how many the numbers jump up by.
- Ask the children to work independently to complete the table of results.
- After the children have had a chance to work through the problem check that everyone has understood the pattern.

Day	1	2	3	4	5	6	7	8	9	10	11
Faces painted	4	6	9	13	18	24	31	39	48	58	69

Drawing together

- Discuss the table of results as a whole class and reach an agreement that Ingrid will have painted 69 faces on the eleventh day.
- Ask the children to work out how many faces Ingrid had painted altogether in the first week (by the end of day 7). (105)

Support

Use smaller numbers, for example, Ingrid painted two faces on day one, three faces on day two and six faces on day three. How many faces did she paint on day five?

Extension

Can the children work out how many faces Ingrid will paint on day 28 if she increases her work rate at the same level?

Further ideas

- Ingrid painted 19 faces on day one, 21 faces on day two, 24 faces on day three and 28 on day four. On which day will she have painted 271 faces? (Day 22.) Alternatively, omit details for some of the early days, but leave the pattern visible.
- Sid Tinker is up to his old tricks again, painting post boxes. On the first day he re-paints three post boxes. On the second day he has painted five, on the third, eight and on the fourth he has painted 12. On which day will he have re-painted 80 post boxes? (Day 12.)

Big apples

Setting the context

The Big Apple Orchard in New York has some of the best and biggest apples in the world. Most apples grown here weigh in at over 1kg. The world record for the largest apple is 1.67kg and every year the Big Apple Orchard tries to beat it.

But this year there's a problem. An army of Scarlet Hairy Caterpillars is back in town and munching through hundreds of Big Apple apples.

When Alex goes apple picking he finds that one out of eight apples has a Scarlet Red Caterpillar drilling a hole to the core. 'I can't sell any of these the caterpillars have had a go at!'

Some pickers are afraid of the enormous red hairy caterpillars. Others aren't so bothered. Henry says, 'Actually, they're alright. The hairs tickle your throat on the way down but really they taste just like apples!'

Problem

How many saleable apples are there out of a crop of 112?

Objectives

To solve mathematical problems or puzzles, recognise and explain patterns and relationships, generalise and predict.
To recognise and extend number sequences, such as the sequence of square numbers, or the sequence of triangular numbers 1, 3, 6, 10, 15.
To recognise multiples up to 10 x 10.

You will need

Paper and pens; weights.

Solving the problem

- Read the story and remind the children how the world's heaviest apple weighs 1.67kg. Can the children think of an object that weighs about this much? A typical bag of sugar weighs 1kg. Weigh out 1.5kg to show the children an approximate measure and compare this to other classroom objects, such as a textbook.
- Ask the children how much six record-breaking apples would weigh (6 x 1.67kg = 10.02kg). How much would a dozen weigh? (20.04kg)
- Read the problem and discuss ideas for ways to approach it.
- Tell the children to focus their minds on the important information in the text. Can they spot superfluous information to discard? For example, although the mass of the apple is interesting, it is not influential to the problem. The crucial information is that one in every eight apples can't be sold.
- Help the children to start the problem by suggesting that they use a table to proceed systematically. Ask: *What should the column headings be?* Encourage pairs to think about the variables at work.
- As a class, start drawing the table and explain that if there are eight apples picked, this is the total; one caterpillar is found in every eight apples, so this completes the apples attacked column; therefore only seven are good, and this number completes the first column.

Saleable apples	Apples attacked	Total number of apples picked
7	1	8

- Explain that counting in multiples of eight will help the children to see how many good apples there are.

- Allow the children time to complete their tables and check to see if the multiples of seven and eight are being increased accurately with each step.
- Ask the children whether it is necessary to complete the table all the way to a total of 112. Have some children realised that they only need to go as far as 56 and then double? Has anyone seen that, if they spot the multiple of seven, they don't need to keep filling in the middle column to solve the problem?
- Look at the completed table together, identifying the patterns.

Saleable apples	Apples attacked	Total number of apples picked
7	1	8
14	2	16
21	3	24
28	4	32
35	5	40
42	6	48
49	7	56

- The children can work out from here that if 56 is doubled, then the apples attacked and the saleable apples also need doubling. So double 7 gives a total of 14 apples attacked and double 49 gives a total of 98 apples fit for the market.

Drawing together

- Place the findings into a clear sentence. For example, for 112 apples picked, 14 were poor quality and 98 were good. Practise saying the answer in this way for other amounts.
- Use the table to ask children a range of other questions to test their understanding. For example, if 10 poor apples were picked, how many apples were picked in total?

Support

- Change the multiples to a simpler figure, for example, *When Alex goes apple picking he finds that one out of four apples has a Red Hairy Caterpillar munching away inside. How many good-quality apples are there out of 24?*
- Start the table off with some missing digits.

Good apples	Apples munched	Total number of apples picked
3	1	
6	2	8
9		12
	4	
15		
18		

Extension

Increase the multiples of apples to include larger numbers. For example, for every 16 apples picked, Alex finds that two have a Red Hairy Caterpillar devouring the inside. If a gross are picked in total, how many good apples are picked? (126)

Further ideas

- If each apple grown weighs 750g, how much do 14 apples weigh?
- Challenge the children to write their own problem story about another piece of fruit that has been attacked by a pest.

The root of the problem

Dear Pong

So, you've found us at last.
What took you so long?
Well, you're here now and so it's time to work. We've made a pretty pattern for you but can you work out how many carrots you'll need to make 25 triangles?

Hope you've eaten plenty of fish because you're going to need all the brain food you can!

The Carrot Crew

You may find the following table helps you.

Number of triangles (t)	Number of carrots (c)
1	3

Fast fingers

If Filly joins two daisies together in the first minute, three in the second minute, four in the third minute and so on, how many daisies will she have in her chain after 15 minutes?

Use this table to help you.

Minute	1	2	3	4	5	6	7	8	9	10	11	12	13	14	15
Added daisies	2						8								
Total	2	5													

Walkie-talkies

Two isosceles triangles set out on a two-week walking holiday...

Use the table below to help you work out the distance.

Day	1													14
Distance travelled (mm)	100													

Flea bag

How many fleas will Jazz have by the eighth week?

Use the following table to help you.

Week	1	2	3	4	5	6	7	8
Fleas	2							

Face painting

What a busy summer painting! I was a bit rusty to start with but I soon quickened up.

On my first day I painted four children's faces, on the second day I painted six faces, on day three, nine faces, and on day four I did 13 faces.

Can you work out many faces I will have painted by the eleventh day?

Use the following table to help you work it out.

Day	1	2	3	4	5	6	7	8	9	10	11
Faces painted	4										

Chapter Three

Diagram problems and visual puzzles

Drawing a problem brings its parts to life and makes it easier for children to understand. Imagining is one thing, but seeing is quite another and so children should be encouraged to sketch out what is being asked. Mathematical marks don't have to be works of art and so you can ease the children's anxieties about having to produce something with line, tone and texture. The trick is to keep it simple.

The problems in this chapter all follow that advice and require little more than a basic outline and some uncomplicated symbols. Sometimes the only marks required are some straight or curved lines. Drawing a picture will also help the children to monitor their progress within a multi-step problem.

Drawing a diagram is a very powerful strategy for communicating a solution because it can help to frame an explanation that might not be readily available in words. A drawing is an ideal starting point to use as a descriptive springboard so that children can find their words, get a feel for the problem and support their reasoning and ideas. By using drawing as a problem-solving tool the teacher is able to see qualities of the children's understanding that might be hidden when using other procedures. Drawings allow expression of feelings and attitudes as well as cognition.

When drawing a diagram, the children will often have to make more than one attempt. This trial and error approach is inevitable as they work through their ideas.

I wish you well

Setting the context

Well, well, well... Where do I start? It all happened so quickly.

There I was, just having a chat to a pound coin and a couple of 10p coins inside a lovely Italian leather purse when someone grabbed me and threw me down a long dark hole. I didn't realise 50ps could fly, but boy did I shift! Talk about a crash landing!

Then I realised where I was, and I wished I wasn't. I was in a well, a deep well but a well without water! It's a good thing there hadn't been any rain lately.

A plaque at the bottom of the well said:

IF YOU'RE DOWN HERE THEN YOU'LL BE DOWN IN THE MOUTH TOO BECAUSE IT'S 19 METRES BACK UP THERE! ENJOY THE CLIMB!

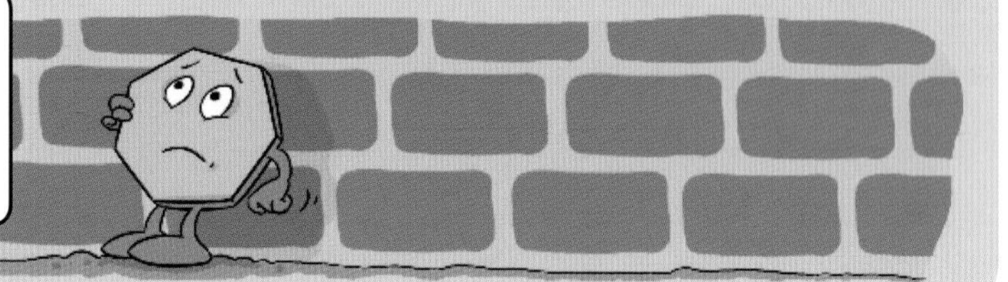

Funny? No. Helpful? Yes – 19 metres was quite a distance, but not impossible.

I started to climb, but the walls were so slimy and greasy I knew I'd have to work hard. I started my escape at 1700 hours. I managed to climb three metres every ten minutes. The problem was, after every three metres I needed a five-minute breather, and I slipped back one metre. My shoes don't have a very good grip you know. Still, I managed to carry on like this and climb out.

Problem

How long did it take the 50p to climb out and at what time did he get out?

Objectives

To explain methods and reasoning, orally and in writing.
To solve mathematical problems or puzzles, recognise and explain patterns and relationships, generalise and predict.
To use all four operations to solve simple word problems involving numbers and quantities based on 'real life', money and measures (including time).

You will need

Centimetre squared paper; rulers; pens; clinometer; beanbags; number line; 50p coin.

Preparation

Measure out 19 metres on the playground or field.

What to do

- Read the story above to the children to set the scene. Revise the 24-hour clock. Ask the children to think about how far 19 metres is. How would they measure 19 metres horizontally and vertically?
- Look back at the story and ask the children to state what the problem is to their maths buddies. Share some of these ideas to see if they recognise that there are two problems being asked.
- Ask the children to tell you what they already know about the problem that will help them to solve it. Can they pull out the relevant figures from the text?
- Challenge the children to think of possible ways of approaching the problem. Emphasise that there will be more than one method.
- Share the children's suggested strategies and write these onto the board or a large sheet of paper for display.
- If none of the children offer it, suggest

that drawing a diagram may help to solve the problem. Ask for the children's thoughts as to how this might be done.

- Now give the children centimetre-squared paper and tell them that 1cm is equal to one metre. Ask them to draw a line so that they can visualise the distance up the well that the 50p has to cover.
- On the board, model the movement of the 50p for the first 20 minutes of the climb, so that the children understand what steps they need to follow.
- Compare the children's diagrams and remind them that the coin would have to go higher than 19 metres to actually move free from the top of the well.
- Take the children outside to demonstrate this in a different way. Using a clinometer, walk and measure 19 metres, asking a child to help you to mark each one metre interval with a beanbag.
- Ask one of the children to 'be' the 50p, moving up and down the well. All together, follow and check the progress made as a whole class.
- Back in the classroom, look over the solution together and answer the original two-part problem.

Solution

Drawing together

- Remind the children that there may be more than one way of solving a problem – although some solutions are more practical than others. Ask what strategy they would select if a playground or field was not available or if it was raining.
- Now encourage the children to visualise the scene and draw it, to bring the problem to life.

Support

Provide the children with a clearly marked number line to demonstrate the progress with a 50p coin.

Extension

Modify the problem so that the coin fell down a 31-metre well. It started climbing at 4am, climbed 2.5m in 15 minutes and slipped down one metre after every five-minute breather.

Further idea

Imagine that the 50p reached 12m, slipped and fell all the way to the bottom. How long would it take the 50p to get to the top now?

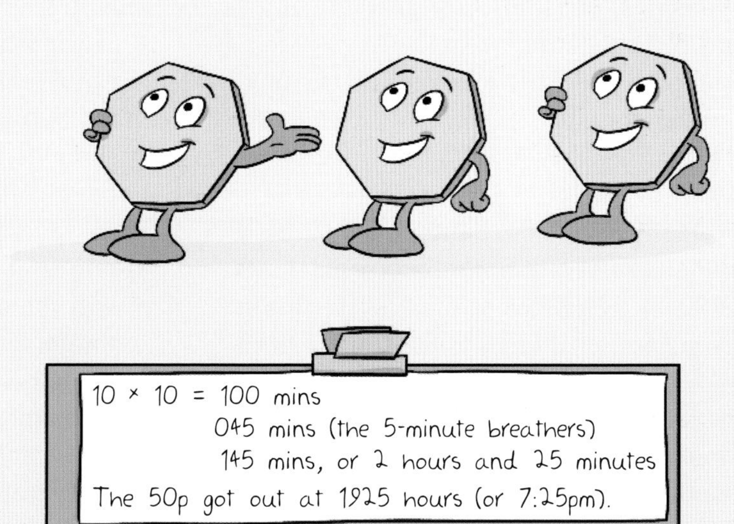

Egging you on

Setting the context

Myrtle McMerryweather has been a battery hen for as long as she can remember. She doesn't like it much. She works long hours in cramped conditions without pay. She'd love to escape, but there are foxes guarding the perimeter fencing of the farm 24 hours a day.

One day she noticed an advert for a competition:

ATTENTION ALL HENS

We are giving one of you the chance to go free range. That's right, free range! If you can solve the following problem you'll be the lucky hen, chauffeur-driven to freedom and a new life in the great outdoors!

Can you lay 18 eggs inside this crate so that each row and column has an even number of eggs in it?

Completed entry forms should be returned no later than lunchtime today. Hens not wishing to participate are welcome to egg on their friends, but we recommend that you don't miss this once-in-a-lifetime opportunity!

Problem

Can you help Myrtle escape from Farmer Moonpie's battery shed? Arrange 18 objects inside a 6 × 4 grid, so that each row and column of the grid contains an even number of eggs.

Objectives

To choose and use appropriate number operations and appropriate ways of calculating (mental, mental with jottings, pencil and paper) to solve problems.

To solve mathematical problems or puzzles, recognise and explain patterns and relationships, generalise and predict. Suggest extensions by asking 'What if...?'

To use, read and write standard metric units and imperial units.

You will need

Egg cartons; cubes or counters; pens; masking tape or chalk; photocopiable page 70.

What to do

- Read the beginning of the story to the children, then re-read the farmer's competition notice.
- Establish what the problem is and write it on the board. Organise the children into mixed-ability tetragon groups and ask them to confirm with each other what the problem is about.
- In their groups, ask the children to think about how they could test out ways of solving the problem. What would we need to help us? Encourage them to come up with inventive suggestions about resources.
- Elicit that drawing is one way to approach the problem, in combination with acting the problem out.
- Provide the children with the photocopiable sheet and cubes or counters. Encourage them to rehearse and sketch their attempts at solving the problem.
- Let the children take their time with the problem, by moving one cube at a time then looking to see what effect this has on rows and columns.
- The children are likely to be challenged and frustrated! So, provide them with clues to push them in the right direction. Point out that one row has to contain six eggs and

three rows have to contain four eggs. Does this help them? Help further by pointing out that there must be three columns of four eggs and three columns of two eggs.

- Freeze the lesson at suitable intervals to share solutions generated so far and to air any frustrations.
- Remind the children to record their attempts on paper for comparing thoughts and previous efforts.

Drawing together

- Share one possible solution with the class and see who else used it. Talk about any other solutions used.

	O		O	O	O
		O	O	O	O
O			O	O	O
O	O	O	O	O	O

- Challenge the children to think whether the same outcome of even rows and columns is possible with an odd number of eggs. Is it possible with any even number of eggs?

Support

Challenge the children to place six eggs into a 3 x 3 box so that there are two eggs in every column and every row.

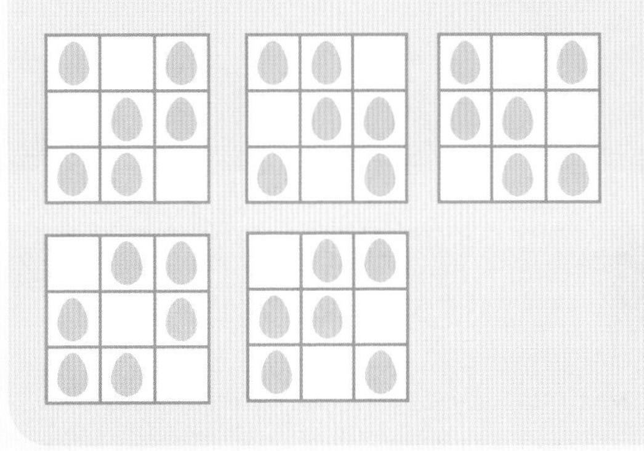

Extension

Say that a certain egg crate can hold 36 eggs. Can Myrtle lay 24 eggs in the crate so that each row and column has an even number of eggs?

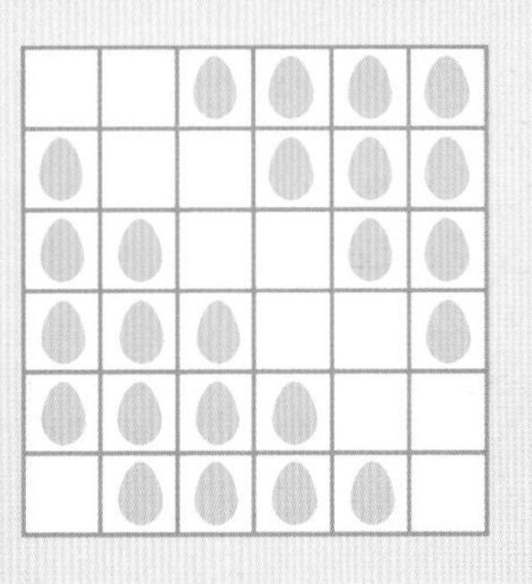

Further ideas

- Set the 'Egging you on' problem using a large 6 x 4 grid in the playground or hall for another class to solve.
- An egg carton can hold four eggs. How many ways can Myrtle fill the carton with blue and green eggs? (There are six different ways: bbbb, bbbg, bbgg, bgbg, bggg, gggg.)

Mountain out of a molehill

Setting the context

To get into the Metropolitan Mole Service wasn't easy. About 1000 moles applied every year for only 50 places. The entrance test included an insect identification exam, a tunnel test and a problem-solving exercise.

Sergeant Star-nosed Moleskin devised a problem he thought would sort out the applicants, 'This will be a whole lot of fun for me. We do get some applicants who make a fuss and say it's too difficult – there are always some who make a mountain out of a molehill.'

All the applicants were nervous and none more so than Mike Burrow. This was his big day and a chance for him to break free from his job at the London Underground digging new tunnels, 'My dad was a plainclothes mole for 20 years and I aim to follow in his footsteps. Let's just hope I've got what it takes!'

Sergeant Star-nosed Moleskin gathered the applicants together in the parade ground and read out the problem, 'Right then you 'orrible lot, your task is this: you are required to dig some mounds in a square shape so that each side of the square has eight mounds. You have 20 minutes from now – so start digging!'

Problem

How many mounds should the moles dig?

Objectives

To solve mathematical problems or puzzles, recognise and explain patterns and relationships, generalise and predict. Suggest extensions by asking 'What if...?'. To explain methods and reasoning, orally and in writing.

You will need

Paper; pens; different-coloured cubes; beanbags/markers; photocopiable page 71.

Preparation

Draw the completed square of molehills ready on the board.

What to do

- Read the story to set the scene, and explain the meaning of the idiomatic phrase 'making a mountain out of a molehill'.

However, don't focus on the problem just yet.

- Begin with some mental maths work, such as how many 50s go into 1000? Can the children express this as a fraction? Decimal? Percentage?
- Now read the problem and make its meaning explicit by writing it on the board. Be sure that everyone understands the word 'mound'.
- Ask the children for their initial responses about what to do. Encourage them to identify the starting points of the problem. Ask: *What do we know and what do we need to find out?*
- Ask the children to think of ways they would find the solution. Would they draw a table? Draw a diagram? Act it out?
- Advise the children to be cautious when solving problems. Remind them not to jump to conclusions. Suggest that if they use more

than one method they can check and support their thoughts.

- Start children off with pen and paper so that they can sketch out their ideas. After a while, provide them with coloured cubes so that they can 'act out' their drawings. A different colour for each side will highlight that corner cubes belong to more than one row.
- Act out the problem systematically using beanbags, markers, or better still children themselves. Stop on the last line for discussion.
- Some children may have thought that the square will have 32 mounds. Point out that the corner mounds are used for both horizontal and vertical sides.
- What did the children notice about the final side? Did they realise that a mound is already used on each corner so they will only need to make an additional six mounds? Look at the solution together.

- Ask the children to think of ways of saying how many mounds there are without using the number 28 itself. For example, 1/3 of 84; a score + 2 tetragons; 50% of 8 × 7 and so on.

Drawing together

Discuss whether the problem can be improved, or RIPEned – Refined, Improved, Polished, Edited. Is there a shorter or more efficient way than drawing?

Support

Use four mounds as a starting point and help the children to build from there. For example: *Sergeant Moleskin wanted the moles to dig some mounds in a square shape so that each side of the square had four mounds.* Model the task in the same way, using drawings and acting out the problem.

Extension

- Increase the level of difficulty so that children have to make a square with ten mounds on each side.
- Change the shape to a rectangle with 28 mounds without specifying the length and breadth.

Further ideas

- If each mound takes 13 seconds to dig, how long will it take to dig the whole square? Give your answer in seconds, and minutes and seconds. (28 × 13 = 364 seconds or six minutes and four seconds.)
- Say that the next task was to make a rectangle shape with a perimeter of 38 mounds. Sergeant Moleskin wants seven more mounds on the longer sides than the shorter. How many mounds are there on each side? (14 on the long, 7 on the short.)

A sore point

Setting the context

Pesta used to love travelling by broomstick. She loved the speed. That, however, was the problem. One day she was caught flying at shoulder height in the local park at over 300mph! The Air Police told her that she would never fly a broomstick again! This broke Pesta's heart. She'd been flying since she was three years old and had never broken the speed limit in her life.

'It's these new brooms. They're much quicker than they used to be. You don't realise the speed you're doing in them,' she sighed. 'Ah, that's it! If I can't fly again, I'll just get rid of the broomsticks... forever!'

Pesta wondered what to do with her collection of eight broomsticks, and thought the best thing for them was to use them as firewood. The broomsticks were too big for the fire so Pesta decided to karate-chop them into smaller sizes so that they would fit.

After a few attempts at karate-chopping the first broomstick, she realised that her hand might break before the stick did. 'My hand is so sore!'

Speaking out loud gave her the sensible idea of finding another method. 'That's it! I'll use a saw! Now I've got one in the fridge I think.'

She did, and so she started sawing. She had worked out that she needed to cut each broomstick into nine pieces. Her first cut took her 40 seconds.

Problem

How long will it take Pesta to cut all of the broomsticks into nine pieces?

Objectives

To solve mathematical problems or puzzles, recognise and explain patterns and relationships, generalise and predict. Suggest extensions by asking 'What if...?'
To explain methods and reasoning, orally and in writing.

You will need

A collection of art straws (or similar); scissors; long cardboard cylinder; pens and paper.

Preparation

Cut the cardboard cylinder into nine pieces.

What to do

- Invite two children to read out the story and set the problem in context.
- Discuss the numbers involved. Are there any numbers that aren't crucial to solving the problem?
- Brainstorm ideas about how to start solving the problem. Do the children discuss superfluous information or are they able to strip the problem to its bare bones and isolate the facts that are important? This is a real skill to develop because narratives can sometimes take children down wrong paths and distract them.
- Ask the children to discuss with their maths buddies how many cuts they think Pesta will make in each broomstick. Thinking about this should prompt them to consider what type of problem this is. Is it a number problem? A drawing problem? Both?
- Ask the children to think of ways that they can bring this problem to life. What resources would they find useful?
- Encourage the children to plan what they are going to do before acting it out, by drawing their ideas on paper first.
- Discuss how many cuts the children think will have to be made to divide the

broomstick into nine pieces. Many children will say nine without thinking, which is why a drawing can be so valuable.

- Provide the children with art straws and scissors and ask them to make the necessary cuts carefully. Encourage them to confirm with each other where to make cuts.
- After the children have had a chance to collaborate and start working on the problem, come together to talk through what needs to be done. Show the children your pre-cut cylinder and count the cuts made. Have the children realised that eight cuts are needed to make nine pieces?
- So, together, work out the total time to cut up one broomstick: 8 × 40 seconds = 320 seconds or 5 minutes 20 seconds.

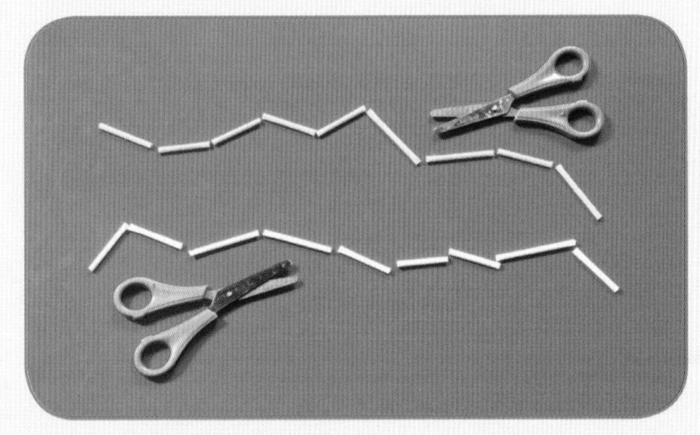

Drawing together

Ask the children to recall how many broomsticks Pesta had in her collection (8). Can they work out how long it will take her to saw all her broomsticks into firewood? (8 broomsticks × 5 minutes 20 seconds is 42 minutes and 40 seconds.)

Support

- Explain that Pesta also had a number of smaller broomsticks from when she was younger. These need to be cut into five pieces each. Each cut takes her 20 seconds. How long will it take her to cut them all?
- Model the solution using an art straw – make sure that the children can see that the number of cuts is always one less than the number of required pieces.

Extension

Ask the children to invent their own similar problem, but using a much longer object. For example, Gerry was sawing up an old telegraph pole for firewood. He has to cut it into 30 pieces. Each cut takes him 37 seconds. How long will it take him to cut the whole telegraph pole?

Further ideas

If Pesta sold her broomstick pieces as firewood at 38p a piece, how much would she get for one broomstick? (38p × 9 = £3.42) For eight broomsticks? (£3.42 × 8 = £27.36)

High five

Setting the context

They just kept on reading the numbers over and over again: '12, 16, 29, 42, 43, 48'. They said them five times over, then jumped for joy, cartwheeled down the street, punched the air and laughed until they cried.

'We're rich, we're rich, we're rich, we're rich, we're rich!' Five times they said it.

No one could quite believe it. The quintet of Inuits from Greenland who had known each other for 60 months won £5 million on the lottery. The date? 5th May. The time? 17:00. Five really was their magic number. They decided that the number five was so special that they moved to the village of Pentagon. They bought a five-acre plot of ice and built five pentagonal igloos on it. The igloos were built fairly close together, but separate from each other. The friends wanted to visit each other often, so a number of ice paths needed to be built too, so that each igloo is connected to every other one.

Problem

How many paths are needed?

Objectives

To solve mathematical problems or puzzles, recognise and explain patterns and relationships, generalise and predict. Suggest extensions by asking 'What if...?'
To explain methods and reasoning, orally and in writing.

You will need

Pens and paper; cubes; art straws; string; photocopiable page 72.

Preparation

Cut enough pieces of string for the whole class. Draw five igloos on the board connected by ten lines as shown in the solutions.

What to do

- Before beginning to find a solution to the problem outlined above, work on the children's knowledge and understanding of the number five. Ask: *What can you tell me about the number five?*
- Here are some five facts: five is the third smallest prime number; five is the only prime to end in five; five is a Fibonacci number because two plus three are five; five is the number of Platonic solids; the Olympics have five interlocked rings as their symbol; the pentathlon is an athletic event with five events; V is the Roman numeral for five...
- Read the scene-setting story and the problem to the children and ask them what a quintet is. *How many years is 60 months? What time is 17:00 in 12-hour clock time? How would you write £5 million in numbers? How big is an acre?* (76% of the size of a football field.)
- Go through the problem text again and ask the children to spot any redundant information. Highlight the important information.
- Hold a thought-shower session with the children and try to come to some agreement about how to solve the problem. Value all ideas and praise efforts that build on previous problems.

- Provide the children with the photocopiable sheet to sketch out their ideas. Allow them to work with maths buddies to bounce ideas off each other.
- Encourage the children to think of the ways that five houses could be arranged. Draw attention to the pentagonal relationship that the quintet share, as this may help.
- Ask the children whether a regular pentagon arrangement connected by a single line would solve the problem. Remind them that each igloo has to be connected to each other igloo.
- Encourage the children to use cubes or other resources to help them visualise the problem further.
- Consider ways of working and debate ideas. For example, how many pieces of string will be needed?
- Organise the children into quintets, using string to act as the ice paths.

Drawing together

- Discuss the number of paths needed to connect each igloo. Did the quintet groups agree?
- Draw five igloos in the style of an irregular pentagon connected by ten lines (see the diagram below).

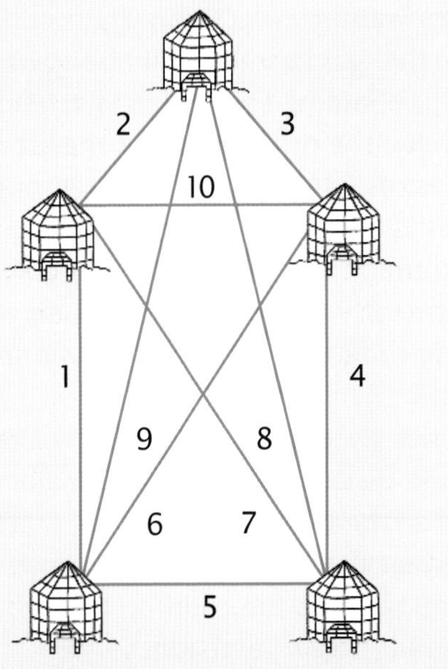

- Show the children your diagram and talk through the solution, path by path. Ask questions such as: *What type of pentagon has been formed? What angles can you spot within the shape?*

Support

- Start with a four igloo problem. Draw a perimeter path connecting the four igloos and focus on the igloos not connected.
- Act out the problem in quartets. Encourage the children to invent their own story based on four friends winning £4 million.

Extension

Suggest that the children tackle similar problems, increasing the number of igloos by one each time. Can they find a pattern?

Further ideas

- Copy the diagram below and set the challenge of connecting each square with the triangle that has the same number (the lines must not cross).

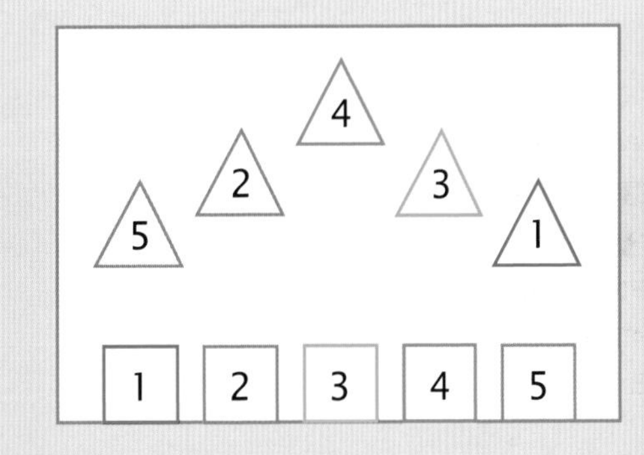

Solution

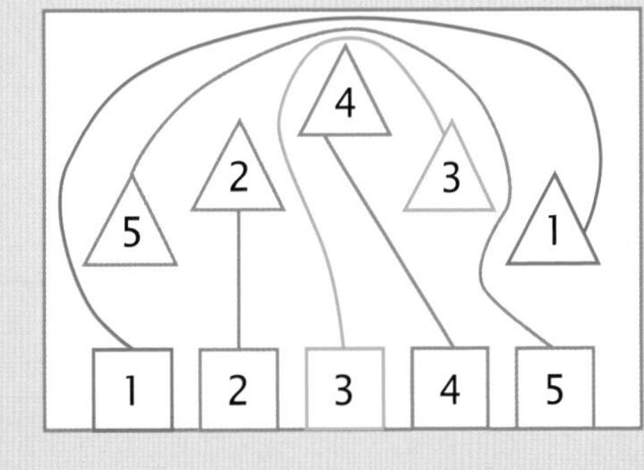

- Challenge the children to invent a story about six friends who won the lottery and bought six houses arranged in a hexagon.

Mexican Wave

Setting the context

Gold medal winners for the fifth time, 'Mexican Wave', the ten-strong gerbil synchronised swimming team, decided to perform their *star* act at the local swimming pool.

'It was the least we could do for all our loyal fans. They've been the real *stars*. We'll do all our old routines and include a couple of new ones as well. We're almost ready.'

Mexican Wave have become the most popular synchronised swimming team in the history of the sport. They have performed all over the world and it would be fair to say they are riding on the crest of a wave. They are renowned for their meticulous preparation and often spend six months of the year just practising one routine. Despite their success, the gerbils haven't forgotten their roots, which is why the people of the small town of Nibble are lucky enough to have these *stars* all to themselves.

'We've just got one routine to finish off, but we're having a bit of trouble getting it together. We're still getting into shape. We might need a bit of help actually! Whatever happens, we'll reach for the *stars!*'

Problem

How do 'Mexican Wave' make a pattern of five rows with each row containing four swimmers?

Objectives

To solve mathematical problems or puzzles, recognise and explain patterns and relationships, generalise and predict. Suggest extensions by asking 'What if...?'. To explain methods and reasoning, orally and in writing.

You will need

Paper and pens; coloured cubes; string; scissors; beanbags/cones; regular pentagons; isosceles triangles; compasses.

Preparation

Draw out the solution on the board.

What to do

- Read out the story, emphasising the words in italic. Then read out the problem and write this clearly on the board for all the children to see and refer to during the lesson.
- Define synchronisation as coordination with respect to time. Encourage the children to talk about what they know of synchronised swimmers. Ask the children to all stand up together after three and perform some simple actions together to illustrate synchronisation.
- Ask the children to think about whether the story contained any clues that might help them solve the problem. Read the story again, asking the children to listen out for any maths shapes.
- Encourage the children to work independently on the problem using a pencil and paper method. Can they draw a five-row pattern?
- Let the children examine each other's ideas and discuss whether they make the desired shape.
- Suggest that using resources such as cubes might help the children to try out the solution in three dimensions.
- Using beanbags or cones, help the children to mark out the problem in the hall or playground. Use string to connect the points together. Discuss starting points as a whole class. Will some beanbags be part of more than one row? Experiment with

possibilities and invite the children to share their thoughts and findings.

- Direct the children towards thinking about star shapes if they find the problem challenging. Draw a couple of sides of a star with the swimmers placed along the sides to show that a swimmer could be in more than one row.

Drawing together

- Point out that a star polygon is the pattern that the swimmers need to make for their show. Show the children an example of a five-point star, also known as a pentacle, pentalpha, pentangle or pentagram. Ask them to think of another name it could be called if they just look at the perimeter lines. Ask: *What shapes make up the star polygon? Are the triangles equilateral or isosceles? What shape is formed in the centre?*
- Show the children the solution on the board. To reinforce and confirm understanding, ask ten children to try to form the pattern in the classroom, using string for the lines.

Support

- Give the children five isosceles triangles and a regular pentagon to make the desired star shape.
- Suggest that the children draw out the solution to the problem using the Logo drawing program.

Extension

- Challenge the children to draw a pentagram using a compass.
- Suggest that the children draw a regular pentagon first and then an isosceles triangles on each edge. Invite them to experiment with varying the lengths of the sides of the triangles.

Further ideas

- Ask the children to think about what other shapes 'Mexican Wave' could form as part of their routine.
- Encourage groups of children to practise and perform their own synchronised routines for a mathematical shape show.

Promotion

Setting the context

Jasper works as a computer game designer in an office block in a town called the Cube. Jasper's job involves a lot of thinking. 'I hate being beaten by a problem. They drive me mad! I can't go home until I've solved it.'

Jasper's hard work doesn't go unnoticed and his boss calls him into the office for a chat.

'Jasper, I'm so impressed with your work that I'd like to offer you a promotion.'

'Wow! That's fantastic!'

'Just one thing. You have to solve a problem to get it.'

'No problem! I love a problem!'

'Well, Jasper, if you can do this, we'll double your wages and give you a company car.'

The boss was confident he wouldn't do it. 'If he does,' he thought to himself, 'I'll eat my hat!'

'Jasper, I want you to turn ten lights on in the office block so that no more than two lights lie in a line in any direction. I'll be standing outside watching!'

Problem

Can Jasper turn on ten lights so that no more than two lights lie in a line in any direction?

Objectives

To solve mathematical problems or puzzles, recognise and explain patterns and relationships, generalise and predict. Suggest extensions by asking 'What if...?'

To explain methods and reasoning, orally and in writing.

You will need

Coloured cubes; masking tape; chalk; beanbags/cones/markers; photocopiable page 73; flipchart paper; paper and pens.

Preparation

Draw the office block diagram on the board. Use masking tape or chalk to mark a large 5 × 5 grid on the floor (big enough for children to stand in). Draw the solution on the board (to reveal later).

What to do

- Read the story to set the scene and ask the children to pick out the important information. What is there that is vital to solving the problem? They will soon realise that the picture of the office block is the key, and the question itself holds all the information required.
- Challenge the children to volunteer any strategies they think might work. Discuss and value all their ideas, however obscure, as there may be alternative ways of working that haven't been considered.
- Fill the first two rows of the office block on the board with the ten lights.

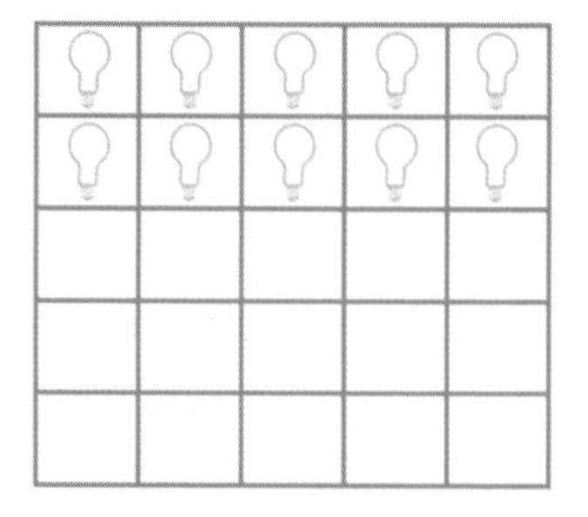

- Ask the children whether the problem has been satisfied. Talk about the difference between rows and columns.
- Provide the children with their own office block grids on the photocopiable sheet so

that they can experiment with different combinations of lights.

- After a period of working, announce a 'freeze' so that the class can talk about their experiences as a whole and reflect on the progress made so far. Has anyone found a particularly successful way of working? Showcase different examples to the class.
- Encourage the children to work systematically by drawing just two lights in a row or column to start with and then to keep adding lights, one row or column at a time.
- Reinforce the children's drawing strategy with an acting-out strategy. Using the grid marked out on the floor in the hall or playground, use markers, or the children themselves to take the part of the lights. Encourage other children to direct the 'lights' where to place themselves. Organise three groups of ten to take turns manoeuvring the lights in the grid, by ensuring that no more than two lights are in a line in any direction. This is a great way to assess the children's spatial awareness and develop their cooperation skills.
- After each group has had a go, record the attempts made on flipchart paper.

Drawing together

- Emphasise that there is more than one way of looking at the problem. Draw out one possible solution on the board and then ask the children to rotate it by 90 degrees.

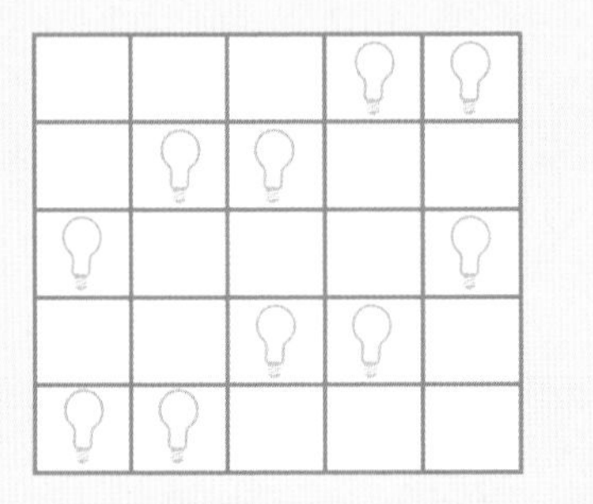

- Point out that rotational symmetry will produce four results (four rotations) that look different but are essentially the same. Remind the children that a figure has rotational symmetry when you can turn it around and fit it exactly on to itself.
- Discuss the solutions and congratulate Jasper on winning his promotion!

Support

Start the children with a 3 × 3 block and three lights so that no light is in the same row or column as any other. Then move onto a 4 × 4 grid and ask the children to turn on seven lights so that no more than two lights lie in a line in any direction.

Extension

Ask the children to turn on 12 lights in a 6 × 6 grid so that no more than two lights lie in a line in any direction; 22 lights in a 8 × 8 grid so that no more than three lights lie in a line in any direction and 38 lights in a 10 × 10 grid so that no more than four lights lie in a line in any direction.

Further idea

Set a challenge for the children to work out how many squares make up Jasper's office block.

A spotty problem

Setting the context

The year is 2090 and archaeologist Professor Sink of the University of Lost Lands has made the most amazing discovery, 'I cannot quite believe it. We were digging away, wondering whether we would find anything – when all of a sudden I saw spots before my eyes. There they were in all their beauty – four spotty rectangles from the Elizabeth II era. What a find! We have a good idea what they are too. We think they are called dominoes and people in the past used to play games with them by joining them together. We are now trying to work out how they may be joined. If we can put them together in some way then they may unlock some secrets of the past!'

Problem

Can you organise the four dominoes in a square so that each side adds up to nine?

Objectives

To solve mathematical problems or puzzles, recognise and explain patterns and relationships, generalise and predict. Suggest extensions by asking 'What if...?'
To explain methods and reasoning, orally and in writing.

You will need

Dominoes; photocopiable page 74; paper and pens.

Preparation

Draw out the solution on the board. Draw the multiplication domino problem on the board ready for 'Drawing together'.

What to do

- Read the problem and give the children a few minutes to reflect on the challenge.
- Ask if anyone has played dominoes. Tell the children that a set of dominoes is made up of 28 rectangles, each having two squares with 0, 1, 2, 3, 4, 5 or 6 spots and every combination is represented. The value of the domino is the sum of the value of the two squares. Tell them that 'Domino' is the French word for a Christian priest's winter hood which was black on the outside and white on the inside. The oldest domino sets date from around AD1120 and appear to be a Chinese invention.
- Ask the children questions related to number properties, such as: *How many of the numbers from one to six are prime numbers? How many are composites (numbers with more than two factors)? What is the largest/smallest number you can make?*
- Ask the children to draw the dominoes in a square shape. Do they see that there is an empty 'window' in the middle?
- Discuss whether keeping one domino in the same place throughout is a sensible strategy. Does this make it easier to try different groupings?
- Give out the photocopiable sheet and encourage the children to draw different domino combinations by trial and error until

a solution has been found.

- Show the solution on the board. Point to each domino in turn, asking the children to tell you what type of number it is and how many factors there are. For example, 5/4: 5 is a prime, 4 is a composite and together they make 9 which has 3 factors (1, 3 and 9).

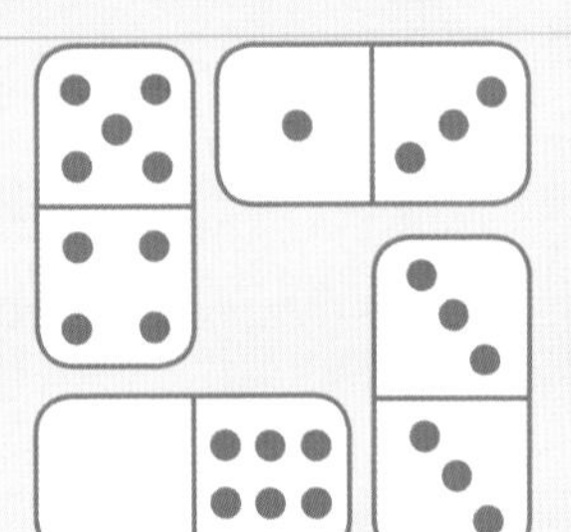

Drawing together

- Recap on the lesson and then provide children with a multiplication teaser. Place dominoes in the pattern of a multiplication calculation.

$6/3 \times \frac{3}{4}$

- Encourage the children to think of other multiplication sums like this, using other dominoes.

Support

- A very useful game to play with a small group is 'Fives and threes'. Play dominoes as usual, but score points when the dominoes at the ends add up to a multiple of 5 or a multiple of 3. Players divide the total on the ends by 5 or 3 and add the answer to their score.
- Alternatively, invite the children to use dominoes to make rectangles (focusing just on the shapes, not the numbers). Can they make a rectangle with just two dominoes? What about three, four, five, six or seven?

Extension

- Use the following dominoes to make a square so that each side has eight dots.
0/0 0/1 1/1 0/2 1/2 2/2 0/3 3/1 3/2 3/3
- The solution is: starting at the top right and going down the page clockwise, draw dominoes in the following pattern 1/1 3/2 0/1 (down right vertical) 2/2 1/2 (bottom horizontal) 0/0 3/3 2/0 (left vertical) 3/1 3/0 (top horizontal).
- Challenge the children to arrange six dominoes so that the total number of spots on the sides are all primes. The sides do not have to total the same, but can they make a pattern that does?

Further idea

Look at the solution and then consider ideas for a new story or a new problem using the dominoes 3/2, 5/4, 4/3, 1/5. Ask the children to work in pairs to think of a different story. Encourage them to think about backdoor maths questions they could ask.

Take your seat

Setting the context

TONIGHT 8PM
A lecture by Professor Filament of the University of Bright Sparks

'How many light bulbs does it take to change a human?'

The atmosphere will be electric!
Admission £5

Rishi had waited all year for this talk. He walked into the lecture theatre to find his seat. Already the place was buzzing with excitement and there were only a couple of people in the room. He was two and a half hours early, but that didn't matter. He couldn't wait any longer. Rishi read the instructions to find his seat:

PLEASE READ THIS CAREFULLY
Your seat is in a row that is fifth from the back and fourth from the front. There are eight seats on your left and two to your right.
WARNING! If you sit in the wrong seat you will be asked to leave the room and someone else will take your place.

After some thought and a bit of searching Rishi said, 'Ah, there I am, right in between those two large women!'

Problem

How many seats are there in the room?

Objectives

To solve mathematical problems or puzzles, recognise and explain patterns and relationships, generalise and predict.
Suggest extensions by asking 'What if...?'
To explain methods and reasoning, orally and in writing.

You will need

Paper and pens; cubes.

Preparation

Draw the seating plan on the board for revealing later in the lesson.

What to do

- Read the question and the problem and make sure that the children understand what the problem is. Some children may jump to the wrong conclusion and think the problem has something to do with the number of light bulbs, or Rishi trying to find where he is sitting.
- Challenge the children to volunteer possible lines of enquiry. Talk about why certain strategies would not be appropriate. For example, ask the children to state explicitly what they are being asked. Does the problem involve looking for a pattern? Can it be worked out through logical reasoning? Is an 'acting it out' method possible?
- Ask the children to close their eyes and visualise the problem as you read the seating instruction again. Read it once more to allow the children to 'see' the room in their mind's eye. Do they find this helpful? Confusing?
- Suggest that drawing a diagram is the most practical method to use. Establish a starting point with the children. If Rishi sits in a row fifth from the back and fourth from the front, does this mean that there are nine rows? Challenge those children who think that is correct.
- Ask the children to work out how many seats there are in a row if there are eight on Rishi's left and two on his right. These numbers may fool some children into thinking there are ten to a row. Demonstrate this with different coloured cubes.

- Ask the children to work independently on the problem so that they can sketch out their ideas before teaming up with a maths mate, to confirm their ideas.
- Remind the children that the problem is to work out how many seats there are altogether, and as part of the challenge they could work out where Rishi is sitting as well.
- When the children have worked on the problem, encourage them to talk about their solutions and discuss any differences in working methods and/or in answers.

Drawing together

- Ask the children to share their solutions as a whole class. Work through any misconceptions before establishing that there are 88 seats (8 rows of 11).
- Show the children the seating plan on the board. Ask them to compare it with their own diagrams.

Support

- Start by concentrating just on the number of rows. For example, if Bertie sits in the seat that is in the third row from the front and fourth row from the back, how many rows are there?
- Act out the problem to cement the idea before drawing. Practise with other scenarios before moving on to seat positions and number of seats.

Extension

Challenge the children with the following: *I sit in a row that is eleventh from the front and eighth from the back. There are 15 people to my left and 9 to my right. How many rows are there, which seat am I sitting in and how many seats are there altogether?* (18 rows, 450 seats.)

Further ideas

- If the lecture theatre was full, how much money did Professor Filament make from his talk? (88 x 5 = £440)
- If the lecture theatre was only a quarter full, how many people turned up? (22)
- If there was a concessionary price for pensioners (£2.75) and 46 turned up, how much did the pensioners pay in total? (£126.50)

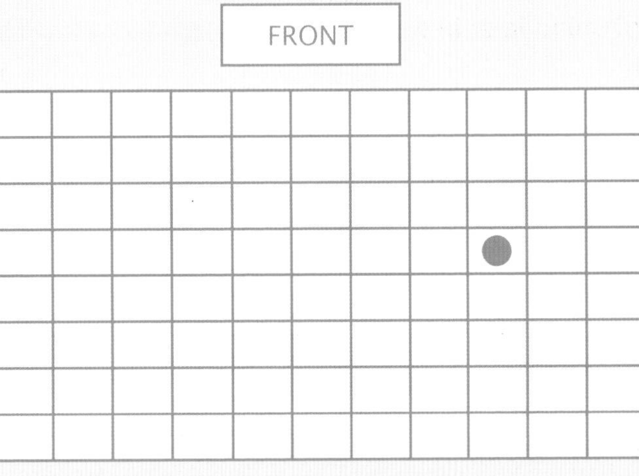

Egging you on

ATTENTION ALL HENS

We are giving one of you the chance to go free range. That's right, free range! If you can solve the following problem you'll be the lucky hen, chauffeur-driven to freedom and a new life in the great outdoors!

Can you lay 18 eggs inside this crate so that each row and column has an even number of eggs in it?

Can you help Myrtle solve the problem and win the competition to gain a free-range life?

Can you show her where to lay 18 eggs inside this egg crate so that each row and column of the crate has an even number of eggs in it?

Mountain out of a molehill

How many mounds should the moles dig?

Use this space to work it out.

High five

Use this space to work it out.

Promotion

The boss called Jasper into his office.

'Jasper, come in! Listen, I've been impressed by your work and I think you are due for a promotion!'

'That's fantastic news! Thank you.'

'However... There's something you've got to do first.'

'Oh, ok, what is it?'

'Can you turn ten lights on in our office block so that no more than two lights lie in a line in any direction?'

'Well, I'll give it my best shot, boss.'

'If you can achieve this Jasper, we'll double your wages and give you a company car!'

Use the following office block grids to help Jasper.

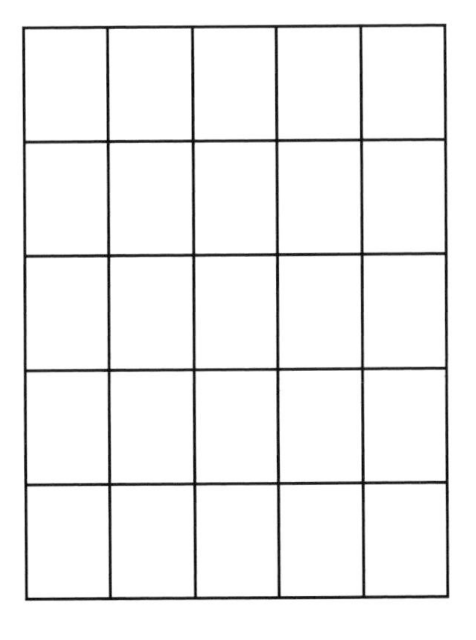

A spotty problem

Chapter Four

Using logical reasoning to solve a problem is like being a detective. It involves following clues, and the clues piece together like a jigsaw puzzle. There's often a lot of hard work that goes into solving a logic problem and there's little room for error.

Logic problems often present a clump of information that needs sorting through and this is best done by drawing a table. Organising information inside a table trains children in systematic working. Encourage the children to read clues carefully and methodically and to use a table to co-ordinate their thoughts in a step-by-step common sense manner.

Presenting the information methodically inside a matrix marked up with symbols provides a visual framework to work from. It helps the problem solver to see the information clearly and the solution is much easier to visualise.

Remember to encourage the children to use the TEAR approach (Think, Explore, Act, Reassess).

Monkeys around!

Setting the context

Five monkeys, Darwin, Gunky, Punky, Marvin and Kong, were invited to a party in a neighbouring jungle and they each decided to take with them a favourite CD to play. They all loved jungle music, but they all also wanted to make sure that their special favourites would be played. They brought music by Banana and the Bashin' Breadnuts, Guava and the Groovy Grasshoppers, Mallow Fruit and the Manic Maypops, Mango and the Mumbo Midges and Pomegranate and the Pipheads.

- Darwin's favourite CD is Pomegranate and the Pipheads.
- Gunky's favourite starts with the same letter as his name.
- Punky and Marvin's groups start with the same letter.
- Punky likes mellow music.

Problem

Can you work out who owns which CD based on the clues above?

Objectives

To explain methods and reasoning, orally and in writing.

To solve mathematical problems or puzzles, recognise and explain patterns and relationships, generalise and predict.

To use all four operations to solve simple word problems involving numbers and quantities based on 'real life', money and measures (including time), using one or more steps, including making simple conversions of pounds to foreign currency and finding simple percentages.

You will need

Five CDs or CD cases marked with the names of the groups; photocopiable page 92; pens; rulers.

Preparation

Draw the empty matrix on the board.

What to do

- Read the story introduction and make sure that everyone can picture the scene before you set the problem. After reading the problem, re-read the clues.
- Lay the CDs out in front of the children. Ask them to tell you, using the given information, which CDs are definitely known as belonging to one of the monkeys (Darwin).
- Organise the children into groups of five to work together to find out another way (apart from physically arranging the CDs) of deciding which CD belongs to whom. Give the children five minutes thinking and discussion time.
- Bring the groups together as a class to share thoughts and strategies. Discuss how solving the problem can be done systematically by using a matrix. Explain the words 'systematically' and 'matrix' as necessary. Teach the children that being systematic means being methodical and working in a step-by-step fashion.
- Show the children the blank matrix/grid on the board and ask them to decide which names will fill the left-hand column and the top row. Ask the children to complete their own matrices on the photocopiable sheet.

● Ask the children to decide what symbols could be used inside the remaining boxes to show whose CD belongs to who. For example, ticks and crosses may be one option but there are many other signs that could be invented. Whatever symbols are used, the key feature is understanding what purpose they serve – to work through the information by a process of elimination.

● Start with the concrete information they know and fill in the box for Gunky using whatever symbol has been decided upon, then set the children to work in their groups to complete the matrix.

● Now come together as a whole class and then complete the grid on the board together, double checking where the ticks and crosses should go.

Support

● Prepare a logic problem using three monkeys and three objects. For example, Fundi, Kinga and Drum meet at the Cactus Café for a drink. They have a banana milkshake, a herbal tea and a mango surprise.

● Use the following clues to work out which drink belongs to who:
Fundi doesn't like fruit.
Drum can't drink dairy products.
Kinga only likes cold drinks.

	Banana and the Bashin' Breadnuts	Guava and the Groovy Grasshoppers	Mango and the Mumbo Midges	Mallow Fruit and the Manic Maypops	Pomegranate and the Pipheads
Darwin	X	X	X	X	✓
Gunky	X	✓	X	X	X
Punky	X	X	✓	X	X
Marvin	X	X	X	✓	X
Kong	✓	X	X	X	X

Drawing together

● Discuss any problems that may have been encountered and recap the importance of a systematic approach. Remind the children that without a systematic method, arriving at a solution would be confusing and just guesswork.

● Explain that the problem is like a jigsaw puzzle and that each piece of information needs to be put together step by step until a solution is found.

Extension

● Challenge the children to write their own logic problems that can be solved using this method.

● Invite the children to think of a problem involving six animals and six objects.

Further ideas

● Bring the problem to life in class by choosing five children to take the part of the monkeys. Label five CDs with the children's favourite groups. Set the problem for another class to solve.

● Use the problem as part of a maths display outside the class for other children around the school to think about.

Brain-knee

Setting the context

It's not easy being the only brain cell in an empty brain. That's why four separate brain cells decided to escape and make a better life for themselves. They contacted each other by telepathy and arranged to meet on the knee of an ancient elephant that lived at Whipsnade Animal Park.

The brain cells are called Axis, Exxo, Lennox and Nixon and the brains are called Leonardo, Archimedes, Einstein and Newton. Here are some facts about the brains and cells:

- Exxo's brain contains four vowels.
- If each letter of the alphabet is given a value so that a = 1, b = 2, and so on, then Axis belongs to Brain Number 95.
- No brain cells belong to a brain whose name begins with the same letter.
- Lennox's brain has a digital root of 1.
- Nixon loves holidays to Sicily.

Problem

Can you work out which brain cell came from which brain? Use the information in the clues to work out the problem.

Objectives

To solve mathematical problems or puzzles, recognise and explain patterns and relationships.

To use all four operations to solve simple word problems.

You will need

Four boxes labelled with the names of the brains; card labels of the brain cell names; cloud shapes to represent the brain cells, one for each group; pens; one A3-size matrix per group.

Preparation

Prepare a matrix on the board.

What to do

- Before the lesson, place the names of the brain cells inside the correct brains.
- Line each of the boxes up in front of the children and explain that they are brain boxes.
- Set the context and then set out the problem. Explain any unfamiliar words and briefly share knowledge and interest about the famous people.
- Organise the children into 'lobes' (parts of the brain) of four or five and give each group a brain cloud to work on.

- Ask the children to brainstorm their ideas about the best way of tackling the problem. Make sure that the children have noted the irrelevance of the zoo setting and so on.
- After five to ten minutes ask the children to pass their brain cloud to another group to share ideas.
- Come together as a class to share ideas from the combined groups. Work out which pieces of information are more helpful than others and try to reach a consensus about the most helpful starting point and the least

helpful place to start. Number the information accordingly. (*Try starting with the clue: If each letter of the alphabet is given a value ....*)

- Explain that a logic matrix is a good way to organise the data given. Show an empty example on the board and fill in the column and row names together.
- Give each group a blank matrix to complete using ticks and crosses to identify the information. Encourage the children to make any notes of problem areas they may encounter, to come back to later when they may be easier to overcome.
- Reconvene as a class to compare responses, then open the brain boxes to reveal the brain cell names.

Support

- Ask the children to work on a similar problem involving three dreams escaping from their owners. The dreams are called Pingi, Pongi and Pungi. Their owners are called Pareen, Paul and Poku. Use the information below to work out which dream belongs to which owner:
- Pingi does not belong to a boy.
- Pongi belongs to an owner with four letters in their name.
- If a =1, b = 2, c = 3 then Pungi belongs to an owner whose letters add up to 50.

	Leonardo	Archimedes	Einstein	Newton
Axis	X	X	✓	X
Exxo	✓	X	X	X
Lennox	X	X	X	✓
Nixon	X	✓	X	X

- Discuss ways of working out the problem. The value of the names (worked out by adding the letter values together) can be done relatively easily: Leonardo = 84, Archimedes = 85, Einstein = 95, Newton = 91. We know that Axis belongs to Brain Number 95 so we can put crosses in all the other names in the same row. We know that Lennox's brain has a digital root of 1 – there is only one brain with a digital root of 1 so Lennox belongs to Newton. Again, place crosses under the other names. This systematic approach, when continued, will enable the children to complete the table.

Drawing together

- Discuss the numbers calculated for each brain name. Check that they are correct and ask whether they think that the brain with the highest number is the most intelligent!
- Recap the meaning of digital root (adding the digits of a number together until you end up with one number).

For example: 95 = 9 + 5 = 14, 1 + 4 = 5

Extension

Encourage the children to invent their own word information clues using the same setting and characters. For example, Nixon's brain contains three letters that aren't symmetrical.

Further idea

Invite the children to work out the numeric values of their names and then find the digital roots. Do they share a value with one of the maths brains?

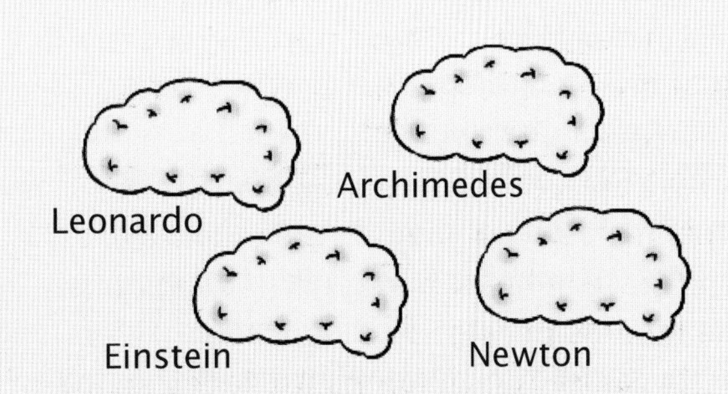

Dog and Bone

Setting the context

I like rings and I love being engaged. I've been engaged more times than I can remember, sometimes for a really long time. It's a great feeling. But things just aren't the same at the moment...

No one has called for a week. Not even a wrong number. I've noticed that this always happens when my owner goes away on holiday. I'm left house-sitting all on my own with no one to talk to.

I was really bored this morning, so I thought I might as well read a bit of the telephone book for something to do. The last thing I read were two pages that added up to 539.

See, I told you I was bored!

Problem

What was the next page number of the telephone book Dog and Bone turned to?

Objectives

To explain methods and reasoning, orally and in writing.

To solve mathematical problems or puzzles, recognise and explain patterns and relationships, generalise and predict.

To use all four operations to solve simple word problems involving numbers and quantities based on 'real life', money and measures (including time), using one or more steps, including making simple conversions of pounds to foreign currency and finding simple percentages.

You will need

Tape recorder; a telephone; a telephone directory; a selection of dictionaries and reading books.

Preparation

Record Dog and Bone's story on tape or CD. Also record the following speech in reply: *Hello, Alexander Graham Bell here. I've enjoyed listening to your ideas and I'm impressed with what you are thinking. I never thought my invention would suffer from loneliness! So, now for the moment of truth. I can reveal that to solve this problem you will need the inverse of multiplication.*

What to do

- Place the telephone on a table in front of the class. Place a telephone book by the side. Play the recording of the telephone talking.
- Read out the problem to the class and make this available throughout the lesson by writing it on the board. Then play the speech from Graham Bell.
- Encourage the children to think of ways that the numbers given in the speech could be used to solve the problem. Centre on the operation that should be used and why. (You have to divide the page number by two.)
- Model the problem with dictionaries and reading books. Start off with a similar example to show the children what is involved. For example, *When Reena opened her book she saw two numbered pages. The sum of these pages was 217. What would the next page be?* (217 divided by 2 = 108.5 – so the page numbers she can see must be 108 and 109. The next page is therefore 110.)
- Allow the children to use the books to put themselves in the place of the telephone

reading through the directory. Give the children thinking time to generate further ideas for similar problems of their own.

- Share ideas as a class and encourage the children to mime using a telephone in order to talk through their reasoning.
- Remember to value all the children's suggestions and display all the responses on the board.
- Tell the children that they are allowed to 'phone a friend' to ask for help with the problem. Pair children into maths buddies so they can talk to each other about how they would solve the problem.
- Say that the inventor of the telephone has been listening and he will lead the class to the correct operation:

In order to solve this problem, divide 539 by 2 to get 269 remainder 1. So one page is 269, the other must therefore be the number added to this to make 539, which is 270. The next page Dog and Bone would read must be page 271.

- Once the division operation has been identified, ask a couple of children to explain the solution.

Drawing together

- Ask the children to comment on the difficulty of the task and tailor further examples accordingly.
- Provide other examples to try in class and for homework.

Support

Work on the main problem together and then provide the children with a problem involving smaller numbers to tackle themselves. For example, the telephone was reading two pages in the directory that added up to 23. What would the next page be?

Extension

Challenge the children to invent and solve a problem about a bored adjective who was reading a dictionary and reached two pages that totalled 3085.

Further ideas

- Encourage the children to think of their own pagination problems to try out on each other or at home.
- Invite the children to set similar style problems to display on the front cover of a book. Ask the children to write out their problem onto an A4 page which could then be turned into a simple book cover with the answer inside. Collate the book covers into a maths problem display.

No place like gnome!

Setting the context

Five Scottish gnomes successfully trained to become astronauts. Their first mission was to go space swimming in the newly found Flintex Solar System in search of the magical wishing fish.

The gnomes all wore the same astronaut swimming uniform of green space costumes and red flippers. The only difference between the gnomes' uniform was their flipper sizes. Gnome flipper sizes are given in letters, and each letter represents a prime number. Size A = 2, B = 3, C = 5 and so on. None of the gnomes have feet bigger than six squared.

'I'm the size of three decagons plus one. They don't call me Big Foot for nothing,' boasted Grog.

'Aye, with a mouth to match! I've always had wee feet. My feet have always been a score minus a heptagon for as long as I can remember,' replied Gron.

'I'm sure mine are smaller than yours. My feet have a digital root of 2. I'm not an even prime size though,' said Greg.

'My feet are an odd pair. One foot is size J minus 12 and the other is the next size up. I need to wear the bigger size,' said Greb.

'If you added 121 to my size you'd get a gross!' Grun shouted.

Problem

Can you use the conversation to work out the flipper size in letters for each gnome?

Objectives

To explain methods and reasoning, orally and in writing.
To solve mathematical problems or puzzles, recognise and explain patterns and relationships, generalise and predict.
To use all four operations to solve simple word problems involving numbers and quantities based on 'real life', money and measures (including time), using one or more steps.
To recognise prime numbers.

You will need

A blank matrix of six rows by twelve columns for each group; name badges/ stickers with the gnomes' names.

Preparation

Write out the letters A to K on card; write out the names of the gnomes onto sticky labels; write the first 12 prime numbers on the board.

What to do

- Read the introduction to set the scene and then set the problem. Re-read the gnomes' conversation to establish the information.
- Taking the children's prompts, note the pertinent information on the board.
- Talk through some of the maths within the problem. For example, ask: *What mathematical sign does a decagon make? How many sides has a heptagon? How many even primes are there?* (Two is the only even prime.) *What is a gross?* (144)

- Ask the children to work in groups to consider possible strategies to solve the problem. Allow plenty of talking and thinking time.
- Debate the strategies volunteered and agree on a systematic solution.
- Emphasise the usefulness of drawing a matrix to arrive at a precise solution and give out a blank matrix to each group. The matrix is a recording tool and helps the children to solve the problem by ordering the information.
- Ask for five volunteers to take on the role of the gnomes and give each gnome a name badge.
- Ask the children to role-play or mime the gnome conversation as you re-read the text. Stop the action at each relevant comment to fill in the matrix as a class. Discuss what a prime number is and fill out the top row with letters and numbers. Revise the meanings of a score and a gross.

Drawing together

- Work backwards through the problem to double-check the matrix has been completed correctly.

Support

Simplify the text so that there are fewer astronauts, and rather than using prime numbers, just use letters. For example, the flipper sizes were F, G, H, I and K. Use the following conversation to help you:

'They don't call me big foot for nothing,' boasted Greg.

'My feet are so wee I can hardly see them!' purred Gron.

'I'm not sure what I am!' said Grog, 'My feet look bigger than Greb's but they look smaller than Grun's.'

Extension

Encourage the children to invent a similar problem based on different-sized space helmets and a different number of gnomes.

	A=2	B=3	C=5	D=7	E=11	F=13	G=17	H=19	I=23	J=29	K=31
Grog	X	X	X	X	X	X	X	X	X	X	✓
Gron	X	X	X	X	X	✓	X	X	X	X	X
Greg	X	X	X	X	✓	X	X	X	X	X	X
Greb	X	X	X	X	X	X	X	✓	X	X	X
Grun	X	X	X	X	X	X	X	X	✓	X	X

- Emphasise the importance of data spotting, reading between the lines and picking up clues along the way when reading a wordy problem.

Further ideas

- Ask the children to consider ways of complicating the problem to make it more challenging.
- Play an odd one out game so that the children get used to recognising prime numbers. Which is the odd one out: 4, 5, 6, 8, 9? (5 because that is the only prime.) 2, 7, 11, 17, 18? (18 because all the others are primes.)

The Pi Place

Setting the context

Four calculators – Napier, Gunter, Shickard and Pascal – are celebrating the end of term at their favourite restaurant, the Pi Place. They love eating maths pies and choose from tomato and tetragons, ham and heptagons, tuna and trigons and onions and octagons.

Pascal: 'I love eating polygons with angles that add up to 180 degrees. I really hope there's a right-angled isosceles one in my pie!'

Napier: 'I'm not too keen on kites and I'm allergic to trapeziums. They bring me out in a terrible rash.'

Gunter: 'I'm not a big fan of those symmetrical pies. I love an uneven heptagon.'

Shickard: 'I really like a pie that's got some fruit in it.'

Problem

Can you work out who chose which pie?

Objectives

To explain methods and reasoning, orally and in writing.
To solve mathematical problems or puzzles, recognise and explain patterns and relationships, generalise and predict.
To use all four operations to solve simple word problems involving numbers and quantities based on 'real life', money and measures (including time), using one or more steps, including making simple conversions of pounds to foreign currency and finding simple percentages.

You will need

A collection of different two-dimensional shapes; pens and paper; photocopiable page 93.

Preparation

Write out the names of the shapes on the board. Draw out a matrix grid to be completed during the lesson.

What to do

- Set the scene by reading out the short introduction. Before reading the calculators' conversation and setting the problem, ask the children to chat with a maths buddy to define each of the shapes:

heptagon – a seven-sided polygon;
isosceles triangle – a triangle with two equal sides;
kite – a quadrilateral with two distinct pairs of adjacent sides that are congruent;
trapezium – a quadrilateral with one pair of parallel sides;
octagon – an eight-sided polygon;
tetragon – a four-sided polygon;
trigon – a three-sided polygon.

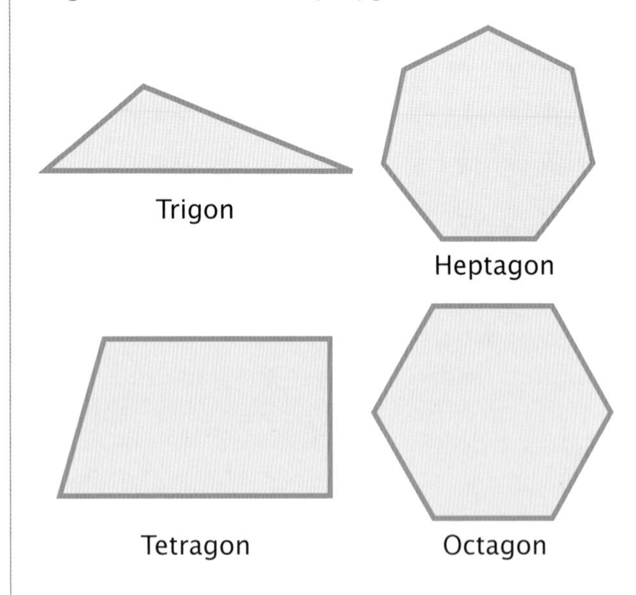

- Ask the children to draw the shapes and then share and discuss their pictures.
- Show the children various examples of tetragons and trigons and test their

understanding of them. For example: *This is a rhombus. Is it a tetragon or a trigon?*

- Read the calculators' speeches (see above) and set the problem.
- Challenge the children to think through possible strategies for solving the problem. Ask them to make notes of their ideas and share suggested plans of action.
- Give the children the photocopiable sheet and point out the empty matrix as a possible systematic approach that could be used. Discuss why this method is appropriate for a systematic approach.
- Organise the children into small groups and ask them to think about what should go in the rows and columns. Relate discussions to any previous experience of using similar grids within other problems.
- Complete the row and column headings together and challenge the children to complete the rest of the grid independently.
- Advise the children to go through each of the calculator's statements and to read between the lines. Can they recognise the clues given and make connections? For example, can they spot that trigon and 180 degrees are connected? Make sure that the children explain and justify their reasoning so that their thinking is explicit.
- Look at the solutions generated:

Support

Concentrate on one type of shape, such as triangles. Use the following as a starting idea. Three trigons – Sindy, Costa and Tani – were celebrating getting 10/10 in a maths test so they went for a pizza at the Angle Café. They chose from egg and equilateral, sausage and scalene and pineapple and right-angle. Who ordered which pizza?

Costa: 'I like a pizza slice with all its sides the same length.'

Tani: 'I love eating a pizza slice with a square corner.'

Sindy: 'I love meat topping on a pizza.'

Extension

Encourage the children to use the same idea, but upgrade the complexity of the problem to include more challenging polygons. Let them craft and create their own conversation clues, for example pepperoni and pentacontagons (50 sides); ham and hectogons (100 sides); cheese and chiliagons (1000 sides) and mushrooms and myriagons (10000 sides).

	tomato and tetragon	tuna and trigon	ham and heptagon	onion and octagon
Napier	X	X	X	✓
Gunter	X	X	✓	X
Pascal	X	✓	X	X
Shickard	✓	X	X	X

Drawing together

- Collect together the children's experiences of solving the problem. Were there any sticking points to overcome? Was the clue about fruit helpful or confusing?
- Ask the children to invent a new scenario involving different mathematical characters and foodstuffs.

Further ideas

- Ask the children to invent a new problem that involves starters and desserts instead of pies. Generate some new names and appropriate clues.
- Which is the most expensive pizza if each letter of the alphabet is worth its numeric equivalent? For example, a = 1p, b = 2p and so on.

Crop circles

Setting the context

'A warm welcome to the 15th Solar System Games. We are delighted to be on Earth again and thank the people of Earth for hosting us. Let's not forget the Moon people too who will be using mirrors to reflect the Sun's light onto Earth so that we have 24-hour sunlight and 24-hour action. Thank you.

The first event of the Games is the Alien Crop Circle Challenge. Will anyone better the circling of Sir Cumference of Saturn? His concentric crop circles won gold last time round, but can he do it again?

You will see the nine fields shown here are separated by hedges. Your challenge is to draw six crop circles of any style so that no three circles are in a line. On your marks, get set... Go!'

Problem

Using a 3 x 3 grid, how can you draw six circles so that no three circles are in a line?

Objectives

To solve mathematical problems or puzzles, recognise and explain patterns and relationships, generalise and predict.
To use all four operations to solve simple word problems involving numbers and quantities based on 'real life', money and measures (including time).

You will need

A 3 x 3 grid for each child; pens; plastic coins; masking tape; images of crop circles.

Preparation

Draw a 3 x 3 field grid on the board. Mark out a 3 x 3 grid on the floor in the classroom or hall.

What to do

- Introduce the 'Games' and set the scene. Talk about crops that have been flattened to form geometric patterns (not always circles). If possible, show the children some examples of the creative designs. Discuss their symmetry and consider how they may have been formed.
- Look at the 3 x 3 grid. Ask the children how many squares they can see. There are more than nine! There are 14.
- Ask the children to think of other ways to describe a square – a polygon, a tetragon, a quadrilateral, a rectangle, a rhombus and a parallelogram.
- Read out the problem and invite the children to work in pairs to discuss how they might solve it. Give the children three minutes silent thinking time to go through their own ideas before they talk with their partner. Then share ideas as a class.
- Hand out grids and coins and point out that drawing circles or moving coins on the grid is one way of solving the problem. Challenge the children to draw their own arrangements of crop circles (just simple circles, not complex patterns).
- When they have done this, call the children together to talk about the arrangements they have made. Name the arrangements after the children that offered them and display them on the board.
- Using a grid on the floor, ask six volunteers to act as the crop circles and work through some arrangements. Ask the rest of the class to contribute ideas.
- Invite the children to comment on what arrangements work and how they would go about finding out more. (Through trial and error.)

- Some children might adopt a guess and check approach, but try to lead them into a more systematic way of working.
- Work through the solution as follows. Because we can't have three crop circles in a line, then how many will there be in a row at most? (Two.) Explain that because we have to draw six crop circles, we have to have two crop circles in each row of the field. Encourage the children to keep two circles in the top row in the same position (for example, top-left and middle) and then vary the position of the other circles underneath, until they find an arrangement that works.

Drawing together

- Look through all the possible answers. Can the children spot two answers that are the same?

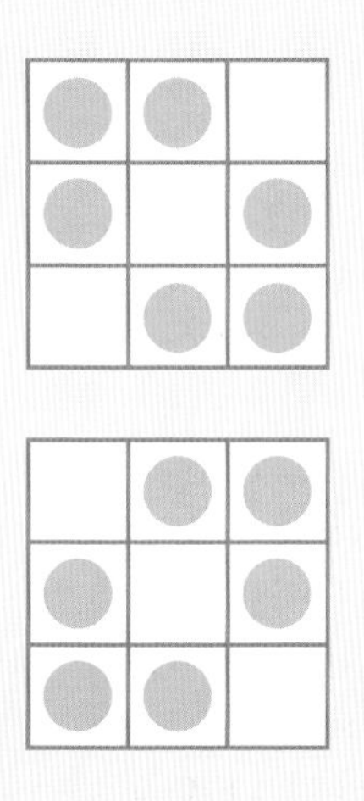

- Essentially there is only one possible solution, although some children may argue that there are two if the solution is rotated through a quarter turn. These are identical and so only one answer can be found.

Support

Ask: *How many ways can you draw three crop circles so that they do form a line?* (There are three ways. It may look like there are four but field one will be a rotation of another.)

Extension

Ask: *For the next round, competitors have to draw five crop circles in the fields so that no three circles are in a line. How many solutions are there?* (There are seven different ways.)

Further idea

In another round of the competition, each alien takes a turn to place a crop circle in a row of nine fields – only one crop circle per field. The winner is the first alien to place three crop circles in a row.

On Cloud Nine

Setting the context

The Oblong twins and the two Pyramid brothers have been visiting friends who live on Cloud Nine where the new Vertex maths theme park has been built. They went on all the rides and are exhausted! They normally take the nitrogen bus but it was fully booked so they hired a magic carpet to take them home to Cloud Cuckoo. Magic carpets are a lot smaller than nitrogen buses so they couldn't all travel together. The carpets can only hold one oblong or two pyramids. They can't afford to hire two carpets but they can ask the same one to take more than one trip. The magic carpet rules are that whenever an oblong goes across, there must be a pyramid on each cloud and the carpet will not fly without any passengers.

Problem

How do they all get home? And how many trips does the carpet have to make?

Objectives

To solve mathematical problems or puzzles, recognise and explain patterns and relationships, generalise and predict.
To use all four operations to solve simple word problems involving numbers and quantities based on 'real life', money and measures (including time).

You will need

Yoghurt pots or similar small containers; cubes of different colours and sizes; photocopiable page 94.

Preparation

Write the solution on the board to reveal later.

What to do

- Read the story and make sure that the children focus on the problem.
- Talk about the shapes in as many ways as you can, throughout the exercise. For example, ask: *What is another name for an oblong?* (A rectangle.) *How many lines of symmetry does it have?* (Two.)
- Ask the children to act out the problem using a container and cubes. Remind them that the magic carpet can only hold so many shapes at a time. Point out that some trips

between clouds may waste time.
- Make sure that the children realise the importance of recording their attempts.
- After a while, if necessary, offer the children the matrix on the photocopiable sheet if they need support.
- Discuss the children's ideas on the number of trips likely to be needed. Keep referring to the properties of the maths shapes and make sure that the children do the same in their explanations.
- Now choose four children to act out the problem in front of the class. Encourage whole-class interaction.
- After a few attempts, try to reach a consensus that both pyramids need to fly first. (Remind the children that whenever an

oblong goes across, there must be a pyramid on each of the clouds.)

- Work through the magic carpet trips, encouraging small groups of children to try it for themselves. Make sure a scribe records the moves.
- Choose two or three groups to demonstrate their interpretation of the problem and discuss each attempt in turn, congratulating all efforts made. How did the children record their results?
- Display one possible way of recording the results as written on the board. For example:

1. The two pyramids fly to Cloud Cuckoo.
2. One pyramid flies back to Cloud Nine.
3. One oblong flies to Cloud Cuckoo.
4. The pyramid on Cloud Cuckoo flies back leaving the oblong there.
5. As for 1.
6. As for 2.
7. The other oblong flies back on his own to Cloud 9.
8. As for 4.
9. As for 1.

Cloud 9	Cloud Cuckoo
OOPP	
OO	+ PP
+ OOP	P
OP	+ OP
+ OPP	O
O	+ OPP
+ OP	OP
P	+ OOP
+ PP	OO
	+ OOPP

(The letters show the positions at the end of each flight. The + sign shows the position of the magic carpet.)

Drawing together

- Share the different ways of working and discuss the chosen methods of recording. Check that everyone understands the sequence and why it works.

Support

- Paurvi works at Zonk Zoo and has to row a zebra, a cheetah and some grass across the Water Hole to the Wild Side. They are all on the Far Side to start with. If she leaves the cheetah with the zebra, the cheetah will eat the zebra. If she leaves the zebra with the grass, the zebra will eat the grass. This will cause a problem on either side of the Water Hole. Can Paurvi get the cheetah, zebra and grass safely across?
- Ask the children to discuss the rules together, such as deciding how many may cross at once. Start the children off on the problem and work through it together.

Extension

Three maths couples, the Litres, the Kilograms and the Metres are on holiday on Co-ordinate Island. The Metres have a yacht moored in Axis harbour which they want to show off to their friends. However, to reach it they have to row a small boat to get there. The boat only holds two people at a time. The problem is the husbands have jealous wives, and none of them will let their husbands be with any of the other wives. How do they all get there?

Further idea

Challenge the children to invent their own problem, varying the numbers of passengers involved.

Musical chairs

Setting the context

'We'd been sat in the waiting room for ages. My brother was getting restless. Now, us hedgehogs can get pretty prickly, you know, when we're bored! There were two rabbits waiting as well, with just an empty chair between us.

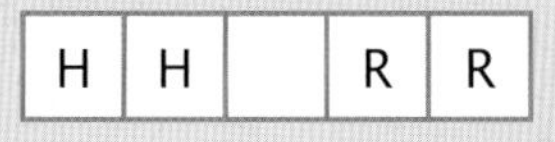

The rabbits were bored too, so we decided to play a game together: musical chairs! Well, there was music playing in the waiting room, so it seemed a shame to waste it on just waiting! The game was basically about swapping places. Here are the rules we played by: if you are a hedgehog you can only move one place to the right or jump over a rabbit. If you are a rabbit, you move in the same way, but to the left. We had such good fun playing this game that we all missed our appointments!'

Problem

Can you work out the least number of moves it will take for the hedgehogs to end up sitting on the right and the rabbits to end up on the left? Note these additional rules:

- **The animals can only move one at a time.**
- **There can only be one animal on a chair at any time.**
- **The animals can slide onto the adjacent chair.**
- **The animals can only jump over one animal.**

Objectives

To explain methods and reasoning, orally and in writing.
To solve mathematical problems or puzzles, recognise and explain patterns and relationships, generalise and predict.
To use all four operations to solve simple word problems involving numbers and quantities based on 'real life', money and measures (including time).

You will need

Large squared paper; coloured cubes; chairs; photocopiable page 95.

Preparation

Have five chairs at the front of the class in a line to replicate the vet's waiting room.

What to do

- Read the story to the children so that they can visualise the scene before you pose the problem.
- Invite the children to volunteer strategies that they think could be used to solve the problem.
- Provide the children with some squared paper and ask them to draw five boxes to replicate a bird's eye view of the chairs. Also, give them coloured cubes to work with: one colour for the hedgehogs, and one for the rabbits.
- Ask the children to record their moves and slides and the total number of moves as they go along. Let them select their own recording methods. The children can use the photocopiable matrix if they need it.
- After working through the problem on paper, act out the problem in 'real life' by using chairs and volunteers. (The animals could be two boys and two girls or children in different coloured tops.)
- Encourage the class to tell you where the hedgehogs and rabbits should move to.
- Ask one of the children to keep count of the moves being made.

- After a couple of attempts, let other children play the part of the animals.
- Now organise the children into groups of four to play musical chairs. Ask the children to think about whether particular animals should be moved in a set order. Remind them to record their moves, slides and total number of moves in a way that suits them.
- The solution can be reached by working logically using a matrix. Eight moves are required as shown. H and R have been used inside the table below but other symbols could be used such as dots and crosses.

H	H		R	R
H		H	R	R
H	R	H		R
H	R	H	R	
H	R		R	H
	R	H	R	H
R		H	R	H
R	R	H		H
R	R		H	H

Drawing together

- Discuss the methods of recording selected and compare how the children noted down their moves. Consider the merits and drawbacks of each method.
- Did the children find acting out the problem easier than doing it mentally or with pen and paper moves?

Support

Work on a similar problem that involves two animals. For example, a giraffe and an elephant:

G		E
	G	E
E	G	
E		G

Extension

Move on to a problem involving six animals, such as three baboons and three gorillas:

B	B	B		G	G	G
B	B		B	G	G	G
B	B	G	B		G	G
B	B	G	B	G		G
B	B	G		G	B	G
B		G	B	G	B	G
	B	G	B	G	B	G
G	B		B	G	B	G
G	B	G	B		B	G
G	B	G	B	G	B	
G	B	G	B	G		B
G	B	G		G	B	B
G		G	B	G	B	B
G	G		B	G	B	B
G	G	G	B		B	B
G	G	G		B	B	B

A total of 15 moves are required.

Further ideas

- Try playing this game with any number of animals, as long as the number of seats is at least one more than the total number of animals.
- Showcase the problem to the school as part of a problem-solving assembly.

Monkeys around!

Darwin, Gunky, Punky, Marvin and Kong grabbed their favourite CDs and swung into party action.

They took with them Banana and the Bashin' Breadnuts, Guava and the Groovy Grasshoppers, Mallow Fruit and the Manic Maypops, Mango and the Mumbo Midges and Pomegranate and the Pipheads.

So, time to be a jungle detective.
Can you work out who owns which CD from these clues?

Darwin's favourite CD is Pomegranate and the Pipheads.

Gunky's favourite starts with the same letter as his name.

Punky and Marvin's groups start with the same letter.

Punky likes mellow music.

Use this matrix below to help you sort out the CDs.

					Pomegranate and the Pipheads
Darwin					

The Pi Place

Napier, Gunter, Shickard and Pascal are celebrating the end of term at the Pi Place. They can choose from the following varieties of maths pie: tomato and tetragons, ham and heptagons, tuna and trigons, and onions and octagons.

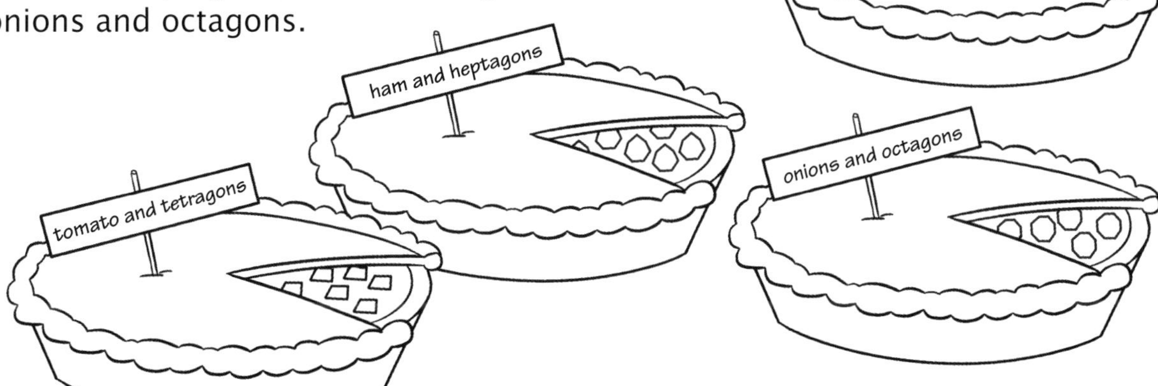

Can you work out who chose which pie based on the information below?

Pascal: 'I love eating polygons with angles that add up to 180 degrees. I hope there's a right-angled isosceles one in my pie!'

Napier: 'I'm not too keen on kites and I'm allergic to trapeziums. They bring me out in a terrible rash.'

Gunter: 'I'm not a big fan of those symmetrical pies. I love an uneven heptagon.'

Shickard: 'I really like a pie that's got some fruit in it.'

Use the following matrix to help you.

	tomato and tetragon		**ham and heptagons**	
Napier				
Gunter	**X**			

On Cloud Nine

Hi, we're the Oblong twins...

...and we're the Pyramid brothers.

We've just been out for the day to visit friends on Cloud Nine. We went on all the rides at Vertex and now we're exhausted. We've hired a magic carpet to get us home to Cloud Cuckoo. Magic carpets can only hold one oblong or two pyramids. We could take two magic carpets but we can't afford to hire two because we spent too much at Vertex! It's going to be a lot of trips!

Remember that magic carpets cannot be flown without a passenger and there must be a pyramid at each cloud if an oblong is going to fly.

How do they all get home?
Use the following table to help you:

O = oblong

P = pyramid

+ = position of the magic carpet

Use this space to make notes.

Cloud Nine	Cloud Cuckoo
OOPP	
OO	+ PP

Musical chairs

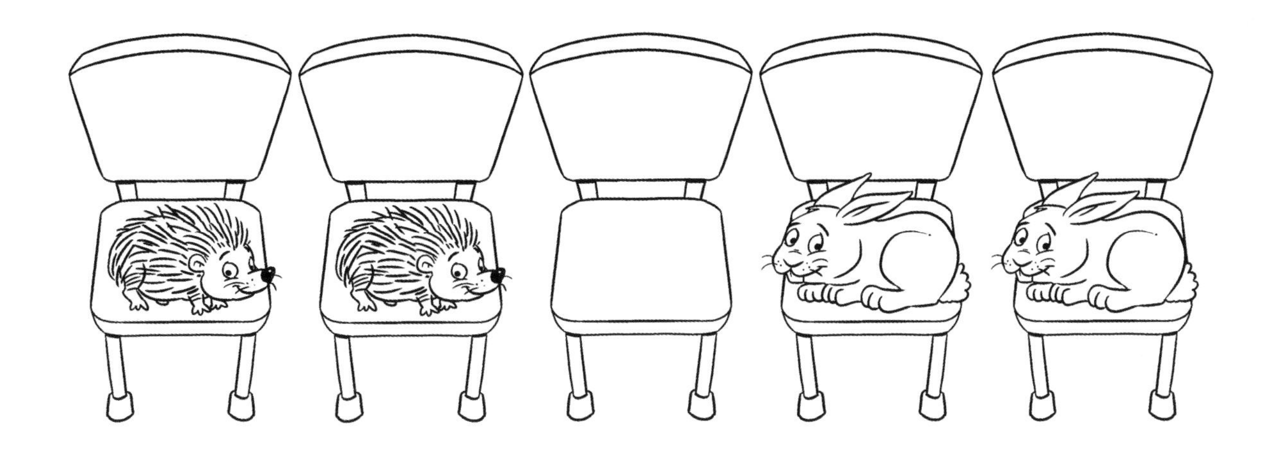

Two hedgehogs and two rabbits were waiting to see the vet. They were waiting so long they decided to play musical chairs to pass the time. They invented some rules: if you are a hedgehog you can only move one place to the right or jump over a rabbit. If you are a rabbit, you move in the same way but to the left.

Can you work out the least number of moves it will take for the hedgehogs to end up sitting on the right and the rabbits to end up on the left?

A table may help you to work it out systematically.

H = hedgehog

R = rabbit

Use this space to make notes.

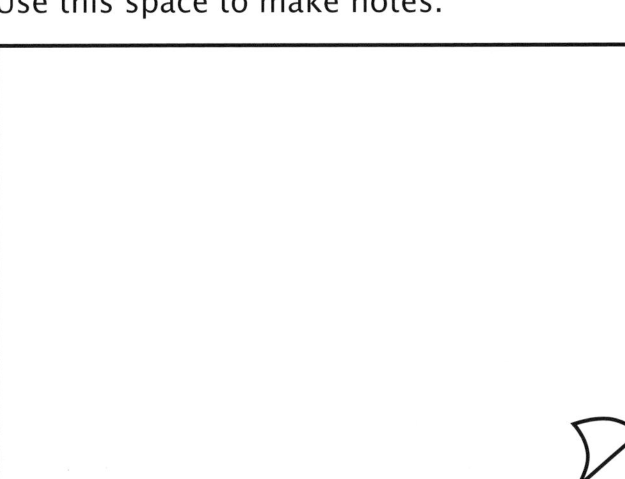

H	H		R	R
H		H	R	R

SCHOLASTIC

In this series:

ISBN 0-439-96556-X
ISBN 978-0439-96556-9

ISBN 0-439-96570-5
ISBN 978-0439-96570-5

Also available:

Shortlisted for the EDUCATIONAL RESOURCES AWARDS 2005

ISBN 0-439-97111-X
ISBN 978-0439-97111-9

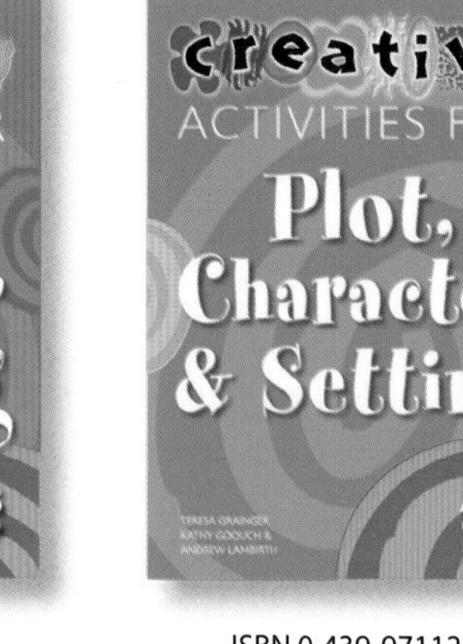

ISBN 0-439-97112-8
ISBN 978-0439-97112-6

ISBN 0-439-97113-6
ISBN 978-0439-97113-3

ISBN 0-439-96526-8
ISBN 978-0439-96526-2

ISBN 0-439-96525-X
ISBN 978-0439-96525-5

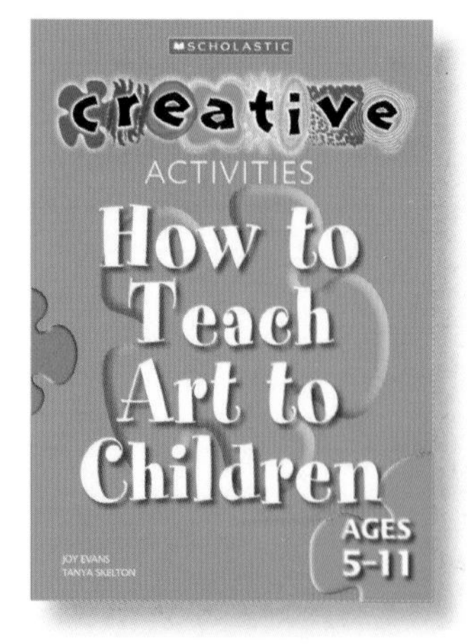

ISBN 0-439-96524-1
ISBN 978-0439-96524-8

To find out more, call: 0845 603 9091
or visit our website www.scholastic.co.uk